SOMERSAULTS IN THE SAND

SOMERSAULTS IN THE SAND

ADVENTURES IN THE GEOLOGICAL MAPPING OF AUSTRALIA

ALASTAIR STEWART

HALSTEAD PRESS

Dedicated to all BMR-AGSO field hands, and to Australia's taxpayers, who gave me a lifetime of exploring, camping, and geological discovery.

WARNING

This book contains a photograph of an indigenous person who is no longer alive.

CANBERRA MMXIV
Published by Halstead Press
Gorman House, Ainslie Avenue
Braddon, Australian Capital Territory, 2612

and

Unit 66, 89–97 Jones Street
Ultimo, New South Wales, 2007

 Designed by megacitydesign.com
Printed in China

National Library of Australia Cataloguing-in-Publication entry

Author: Stewart, Alastair, 1939- author.

Title: Somersaults in the sand : adventures in the geological mapping of Australia / Alastair Stewart.

ISBN: 9781925043037 (paperback)

Subjects: Stewart, Alastair, 1939-
Cartographers—Australia—Biography.
Geologists—Australia—Biography.
Geological mapping—Australia.

Dewey Number: 559.4092

CONTENTS

ACKNOWLEDGEMENTS

The remote dry nature of Australia's outback results in many odd happenings, some of them funny, some dangerous, and some tragic. In 1990, Larry Harrington of BMR asked some of the 'old BMR mappers' to write down their fieldwork experiences, which I did. Many years later, Lesley Vincent encouraged me to turn the stories into a book. Lesley, Diana Pillinger, my daughter Alexandra, Evelyn and Warwick Nicholas, Dean Hoatson, Colin Watson, Matthew Richardson, and Ray Smith reviewed and improved the manuscript. Alexandra drew the sketches, and Joe Mifsud drew the locality maps. Others who helped were Stewart Cole, Diane Constans, Guy Featherstone, Erwin Feeken, Stephen Hodgkin, Lis Hoorweg, Songfa Liu, Rae Luckie, Ken Mitchell, Malcolm Nicoll, Lyall Offe, Euan Ritchie, Barbara Triggs, Allan Wells, Giulio Zambon, and the Manuscript Appraisal Agency. Some of these stories first appeared in the Earth Sciences History Group Newsletter of the Geological Society of Australia, and are reprinted by permission.

Finally, I thank Evana Ho for her invaluable contribution; she reviewed, edited, and greatly improved the entire manuscript.

FOREWORD

This amusing account of fieldwork in the outback should be of interest to anyone who has travelled there, or is contemplating a journey there. The humorous anecdotes belie the serious and important nature of the work carried out in the geological mapping program of Australia by geologists of the Bureau of Mineral Resources. Their work, and the splendid regional and detailed maps made under sometimes difficult conditions, have formed the basis for most mineral exploration activity in Australia. The organisational work of party leaders was impressive; they had to not only plan and carry out the geological survey, but also arrange the field vehicles, sometimes including helicopters in the early days, cooks, field hands, draughtspersons, and visitors. Some brought their families to the field camps to avoid long separations. Occasionally some of the support staff found it difficult to work in the isolated conditions, and that made difficulties in finding local replacements.

I first met Alastair in 1967 on an excursion that BMR organised to show geological academics their work in central Australia. He woke the whole camp up every morning by smashing flat the burnt out food tins from the night's dinner. This was done so that they would rust away quicker on burial. This introduction led several University Departments to attach graduate students to the BMR field parties for their degree work—in my case, two in the Centre and two at Mt Isa. The BMR support was invaluable, both for the geology and for the student's introduction to working in the outback. On publication the thesis work enhanced the scientific results of the surveys.

All the party leaders have retired now, and the nature of the present scope of work by Geoscience Australia has changed, so that such surveys have become part of the nation's history.

Mike Rickard
Australian National University

1

THE START

I'm one of the lucky few who get paid for doing something they love. I'm a geologist. However the path that got me here was not always smoothly paved. I will tell the story from the beginning.

My mother and father met in Germany after the end of World War I. She was visiting from England to meet her German relations, and he was a young cavalry officer in the Australian Army. They married in Australia in 1922, when she was 17 and desperate to get away from her stepmother, whom she hated. As she matured, they grew apart, she being a liberated extrovert and he an introverted perfectionist. But they lived together until the start of World War II, when my father went off to fight in Greece.

Colonel Alan James Stewart, OBE, the author's father (front, middle), with four officers of 1st Armoured Brigade Company, Royal Australian Army Service Corps, Singleton Camp, NSW 1950. Unkown Australian photographer.

For my first six years, Mum brought me up single handed. When the war ended, she set up a Box and Cox arrangement – my Dad went to live in the army barracks in Sydney while she and I stayed in Melbourne, and a tall, handsome ex-Air Force

Margery Stewart, the author's mother, and Ronald MacGregor at home in Beaumaris, 1957.

policeman, Ronald MacGregor, came to live with us. Mum reckoned he looked like Clark Gable and James Stewart rolled into one. He used to move out for a few weeks each year when my Dad came home on leave, and then moved back in again when Dad returned to Sydney. MacGregor became my de facto stepfather, and I was very fond of him. After my Dad retired from the army in 1953, Mum moved out and set up house permanently with MacGregor. Three years later, Dad moved to Queensland and Mum and MacGregor moved back in with me. My parents never divorced, and as far as I know, my two fathers never met.

As a boy I found scientific things fascinating. I read and re-read books that Mum gave me, such as *The Conquest of Space*, *Men of Other Planets*, and *The Earth and its Mysteries*. I made plaster of Paris models of the planets, although I never worked out how to put the rings around Saturn, without touching the planet. Mum gave me a chemistry set, and I once gave her a marvellous fright showing her the power of a homemade explosive. I even read parts of a textbook that she gave me called *General Astronomy*.

In my last year at preparatory school, our science teacher, Mr A.J.P. Kelso, brought some rocks in to show us, including a piece of 'rotten' or weathered crumbly granite. I can still see those glinting black mica flakes, rusty yellow feldspar grains, and cloudy quartz crystals. We lived in the suburb of Beaumaris, which has a cliffed coastline cut in red-brown sandstone on the eastern side of Port Phillip Bay, and earlier that year I had found at the base of the cliff at the bottom of our street a symmetrical heart-shaped pebble etched all over with many small rings and holes. I took it to class and showed it to Mr Kelso. Straightaway, he identified it as a fossilised sea urchin, *Lovenia forbesi*, which had lived about 23 million years ago.

My senior school was strong in physics, chemistry, mathematics, and languages, and almost all the teachers were enthusiastic and inspiring. However, the school did not teach biology, and when it did start teaching geology in my final year, they taught it only to boys who were hopeless at maths. The physics and chemistry 'swots', including myself, naturally looked down on geology. The next year I started a science degree at Melbourne University, and for the first year we had to take physics, chemistry, pure maths, and an elective of applied maths, zoology, botany or geology. I chose geology, to see what it was like. I enjoyed it very much, especially the field trips, where a horde of us walked over fields and parks in and around Melbourne, and we discovered some of the active processes that go on in the Earth. One of the field trips was to the cliffs at Beaumaris where I had found the fossilised sea urchin, and farther down the coast near Frankston on this same trip, we collected perfectly preserved spiny marine snail shells and crystals of pyrite from a cliff of soft grey clay.

My other favourite pastime when growing up had been scouting, and I went on many overnight hikes and camps in the bush and farmland east of Melbourne. By the end of my first year, when it was time to decide what to major in, the joy of combining the study of rocks with outdoor living saw me settle on geology.

After I graduated from Melbourne University, Mum said 'Go into the oil business.' Just like that. I'm not sure how she knew about oil exploration. For many years she worked at Myer's lending library, where the librarians' job was to suggest to borrowers the titles of books they might like. So she must have known the jobs, interests, likes and dislikes of a great many people.

When I agreed, she talked to some of her contacts, and Dudley Pilcher, a Melbourne businessman with a craggy kindly face and the coldest hands I have ever touched, told me to talk to the Bureau of Mineral Resources (BMR, renamed Australian

The author at graduation, Melbourne University, 1962.

Geological Survey Organisation, AGSO, in 1992, and Geoscience Australia in 2001) in Collins Street. BMR was part of the Commonwealth Department of National Development. I went there, and talked to a senior geophysicist, Mr Bob Thyer. He told me to write to BMR's Geological Branch in Canberra. To my amazement, a reply came from the Chief Geologist, Dr Norman Fisher, offering me a job, and I accepted. So on 1 April 1962, at the age of 23 and almost wholly lacking in experience, I drove from Melbourne to Canberra to start out as a geologist.

2

FIRST TIME IN THE RED CENTRE

(1962)

On a cold, clear, sunny Canberra morning at 9 o'clock, I reported to Fred Gocentas, the BMR staff clerk in the MLC building on London Circuit. Fred was a navigator on weekend car rallies, and at work, friendly, helpful, and direct. After I'd signed on as a Geologist Class 1 (Temporary) at £1500 a year, he gently pointed out that the public service working day started at 8:30.

We walked to the BMR offices on Childers Street, about ten minutes away—a row of grey fibro-cement huts of different lengths. In one of the longest huts, we went down a brown linoleum corridor and into one of many small offices to meet my party leader, David Forman, a tall, dark, good-looking Western Australian in his late 20s, with a tanned complexion and a reserved manner.

Beside the door was a floor-to-ceiling stack of grey metal shelves, filled with small grey cardboard boxes containing chunks of grey and red rock. Bookshelves in the room were of the same grey metal. The grey huts, brown lino, grey shelves, and grey sample boxes made a rather depressing introduction to BMR.

With a few weeks to fill in before we left for the field, my first job was to hand colour three copies of the two geological maps produced from the previous year's fieldwork. Peter Cook, who had joined several months earlier from England, coloured another three. The maps were about 1 metre by 75 centimetres, and each one took several days to colour.

We used government-supplied Derwent D19 pencils imported from Britain. Over the years we got to know these pencils so well that we sometimes used the numbers in conversation.

'What colour hair did she have?'

'Sort of a mousey brown, about Derwent pencil 57.'

The coloured maps were designed to accompany copies of a report on fieldwork that went to the BMR top brass—Director, Chief Geologist, etc. I kept to myself the thought that this humdrum job was a bit beneath me. Just as well, since I would be colouring many more maps over the next 30 years. Colouring by hand was the best way to find drafting errors in a map, and it also taught us the extent and distribution of rock units in areas we had not mapped ourselves.

I felt rather nervous about the impending field season. I knew nothing about life and work in the outback. The sample rocks on the shelves looked almost identical to each other, greatly weathered and mostly red. 'How am I ever going to find these interesting? Will I even be able to identify these strange rocks in the field?'

Our field party, the Bloods Range Party, had to map about 15,000 square kilometres of the Amadeus Basin in the southwest of the Northern Territory. In 1874, the explorer Ernest Giles had gone there and named Bloods Range and the Petermann Ranges. Now the area had potential for petroleum and metals.

Our party assembled in Adelaide in late May: Dave Forman and myself, two field assistants, a cook, and a mechanic. Dave had done several field seasons in the eastern part of the Territory, but this was his first season as party leader. Ian Hughes and George Kucinskas were the field assistants, Jim Kelly the mechanic, and Max Mansfield was cook. For Max and myself it was our first time out in a field party. I was the 'greenhorn' in the group.

Field parties took a lot of organisation, and the most important thing was to select good staff. They were hired only a short time before the start of the field season, and suddenly, a group of strangers had to learn to live for several months with each other. The party leader needed skill and responsibility in interviewing and choosing, and wasn't always successful. This time, Dave had chosen four excellent field hands.

A field assistant's main tasks were to drive the vehicle, fix punctures while the geologist examined the outcrop, get the vehicle going after breakdowns,

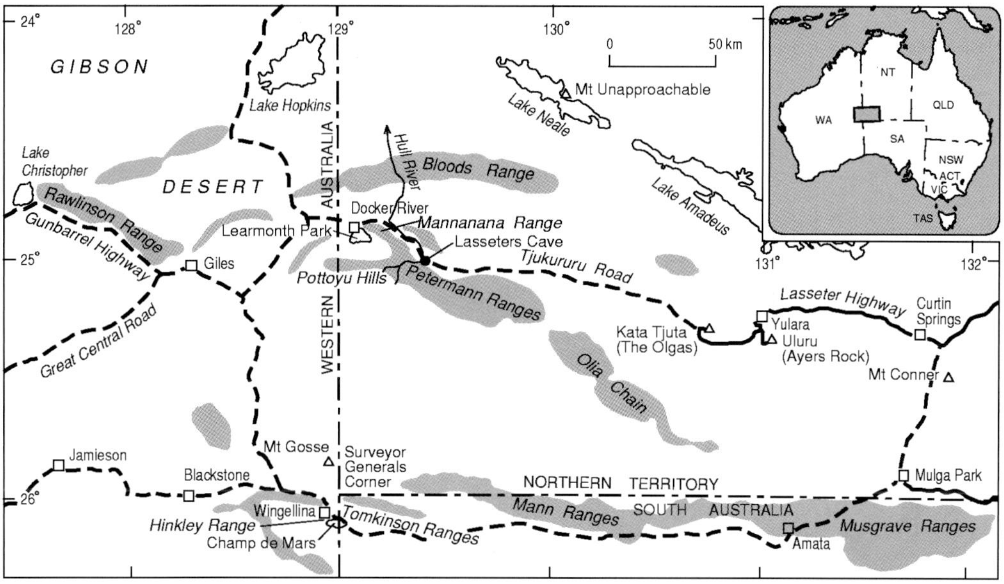

Locality map, southwestern Northern Territory and environs.

and do most of the cooking in the evening after making camp. In base camp, they refuelled and restocked vehicles for the next week's work, and where the party was camped within reach of a town, sometimes went on a supply run, generally for fuel, mail, vehicle parts, and bottled gas.

The mechanic's job was to service and repair the vehicles. This occupied him for the whole weekend, but often left him with little to do during the week, so he usually doubled as a field assistant and went out on regular traverses with the geologists. That could leave him with little or no free time, and it sometimes required creative time sheet entries to pay mechanics for the extra hours they worked.

The cook cooked for everyone in base camp, and on weekends baked bread or damper, unless the camp was within reach of a bakery. The job could be stressful—during the week there might be only a draftsman to cook for, and then on Friday evening, four to eight or even ten people would come in expecting decent food all weekend plus fresh bread for the Monday morning. Cooks often left during a field season, and then came the problem of finding a replacement, usually through the Commonwealth Employment Service in the nearest town. The calibre of candidates in country towns looking for cooking jobs was usually well below what was required for a field party.

Some good field staff used to sign on and come out each year, and alternate with other work between field seasons. There were field hands who could

Jim Kelly wiring up broken U-bolt on Landrover, with David Forman looking on, near Lake Neale, 1962.

assemble a party, collect stores, take the convoy out to the mapping area, set up the base camp, and then collect the geologists at the nearest airport when all was ready for traverse work to start.

We collected vehicles and stores at the Department of Supply No. 2 Store at Finsbury, on Adelaide's Old Port Road. This was one of three huge, cold and windy, galvanised iron sheds like hangars, filled to the roof with shelves of camping gear and tools that government field parties needed.

The storemen were a fascinating mixture of crotchety middle-aged men who had no interest in scientific work. Most of them cared nothing for our desire to leave Adelaide as soon as possible, and were unbending in their procedures and fixity of purpose in ensuring that all five copies of every sheet of the stores requisition order were properly filled in. They sometimes even talked like a stores list. I once heard one—Jim Nader—walk past muttering 'Boxes, tucker; boxes, tucker' under his breath. They had little idea of what working in the outback entailed, and often did not see, nor care, why we needed certain items, such as small shovels rather than large ones for digging under a bogged vehicle, or two jacks per vehicle to replace a broken wheel spring. Over the years as we got to know them, particularly Tom Webb and Jim Nader, and they got to know us, they became more friendly and cooperative. I found later with my own parties that if we specified exactly what we needed, added a comment as to why for out-of-the-ordinary items, and did so well in advance, then the stores issue became easier and even a little quicker.

I suppose the life of a storeman has few rewards, apart from regular hours and salary. They issue equipment which has no meaning and certainly no significance for them, and months later receive much of it back, worn, dirty, and some of it damaged. They get it cleaned, repaired, or replaced, and then issue it again. They seem to produce nothing tangible, and yet their work is essential.

In the mid-1980s, when the Department of Supply had ceased issuing equipment for BMR, BMR's own store in Canberra took over that role, and had the excellent idea of sending a young storeman, Shane Nancarrow, out with a field party for a week to see how the equipment was used and how field work was done.

Shane had an extraordinary ability. All government stores items have a number, mostly a six digit combination of letters and figures. All the numbers had to be written on the order for all items required by a field party setting out, and again returning the stores when they got back. Shane knew the numbers off by heart, and could write them straight down on the forms instead of having to look them up one by one.

I took Shane on a three day traverse in the Mount Isa area. He came along on all my cross-country walks, was interested in the geology, and asked many questions. When the time came to return to the field the next year, the attitude of the BMR storemen had improved markedly, and we had good relations with them from then on.

The government supplied all our camping gear, but we paid for all the food, so one of the geologists had to act as messing officer for the party. Because food was cheaper in the capital cities than in outback towns, we bought the food in bulk and took it with us. In a large party going out for several months, members had to provide quite a lot of money, which often they did not have, particularly the field hands, who were just starting a new job. So Treasury used to advance the necessary money and then deduct a portion from each person's fortnightly wages. Sometime in the 1970s, after this practice ceased, BMR began advancing a large proportion of the travelling allowance for the field season to each party member. This worked well except for the time when a new field hand accepted his advance of several thousand dollars, and was never seen again.

The vehicles we collected, all four wheel drive, all brand new, included a 3 ton Bedford truck, a 1 ton International truck, two short wheelbase Landrovers, and a two wheel box trailer. All except the Bedford were painted 'Limestone

Yellow', as written on the inside of their doors. This strikes me as quite funny now, a geological colour for our geological vehicles.

On collecting the Bedford, Jim was amazed to find that the engine was filled to the brim with oil. We speculated that it must have been poured in to protect the engine while it was being shipped to Australia. Jim had to drain out quite a lot of the oil before the engine would start.

Finally, after four or five days, we were able to get away from the Adelaide store and depart for the Northern Territory. Max and Jim led the convoy in the Bedford, it being the slowest vehicle and also greatly overloaded, carrying eight tons instead of the designated three.

After going a few kilometres along the Great North Road, the Bedford stopped. Several spoonfuls of sand and gravel were blocking the fuel supply, and Jim fixed it by simply unclipping the bowl and tipping the sediment out.

After a few kilometres more, the truck stopped again. It was only 9.30 in the morning, but it couldn't drive past the first pub out of Adelaide. By the time we arrived at Port Augusta it was evening. We drove a few more kilometres north, and then pulled over a short distance off the road and camped in the dark in a small cleared area.

This was my first experience with a BMR swag and stretcher. The stretcher was new, stiff, and had a mind of its own—the metal rods for the sides had a telescopic join halfway along, which needed strong arms and fingers to put together. The spring-steel legs were worse—they had a wide and shallow W-shape, which had to be squeezed by bearing down with all of one's weight so that they could be fitted over lugs on the side rods, all in the pitch dark. Eventually I settled into my swag, and then Ian came over and told me, 'Always sleep with your feet into the wind, mate.'

There is nothing worse than the wind blowing over your head and shoulders when you're trying to sleep. I had to wriggle out, put my boots back on, lick a finger and hold it up, work out where the wind was blowing from, drag the stretcher and swag around, climb back in, and at last go to sleep. We got a nasty surprise just before dawn, when a station vehicle drove noisily through our camp right between us still in our swags—we had camped on a dirt side road.

Ian was a kind-hearted chap, and we got on very well and much enjoyed going on traverse together. He was worldly and knowledgeable, and had had a variety of jobs, one of which was as a lighting designer. He could design anything from illuminated shop signs to lighting for offices and factories, and had been a field hand with a CSIRO biological party in the Top End of the Northern Territory the year before coming out with us. After the BMR field season ended, he completed a PhD and then taught at the Australian National University Geography Department.

Ladies found him irresistible, and he had several long-lasting relationships and sired more than one crop of children. Most of us grew beards during the field season, and Ian's was the most remarkable as his beard and moustache were distinctly and unmistakably Asian in appearance—he looked like Genghis Khan until six weeks into the field season he shaved them off.

Max was a dour non-committal type, good hearted, hard working, but with a rather bitter outlook. He became disillusioned as the season went on, and was nearly sacked in the last week of the field season.

/ / /

We arrived at the security gate of the Woomera Rocket Range at about midday on the second day. We had a permit to drive through the range instead of around it, and we arrived at Woomera township just in time to have lunch with Dave's sister Anne and his brother-in-law, Paris Drake-Brockman, a senior army engineer at the rocket range. Woomera is one of those single purpose towns that are found at mines, communications bases, and in this case, at a military base for testing rockets. Its modern houses made an extraordinary contrast with the stony desert around it. I felt a bit out of place walking unwashed and unshaven into a strange suburban sitting room.

My Dad's great friend, Colonel Richard Durance, a former deputy at the Army's Department of Supply and Transport, was the base commander, so I took the opportunity to ring him up and say hello. He immediately came over and joined us, and then took me away for lunch at his place, where he made fun of my day old beard. Still, I let it grow for the rest of the field season. Anne doted on her brother, and much enjoyed introducing Dave to the Colonel.

/ / /

The trip from Adelaide to the Northern Territory took several days along the Stuart Highway—a narrow corrugated strip of dirt and stones or, elsewhere, sand. We drove through the opal-mining town of Coober Pedy during a thick dust storm, with the visibility down to a few metres. Most of the town had been built underground to avoid the hot summers, which can be truly debilitating in central Australia, as I was destined to find out. Only two buildings were visible through the gloom, and the highway, which was just wheel tracks in the dirt, turned sharp left between them. We did likewise, and continued on through the opal diggings without seeing a single person.

Farther north still, in sand dunes, the road—which was wheel tracks in the sand—was deserted. We stopped for lunch one day, and simply left the convoy filling the entire road; I felt a bit worried by this, but not a single vehicle appeared the whole time. At Kulgera, just over the border in the Northern Territory, we turned west and headed through Victory Downs to Mulga Park and then northwards to Curtin Springs. We camped in the dark by the road north of Mulga Park, and woke to a clear frosty morning and a magnificent view of the flat-topped Mount Conner in the dawn light.

Camping at Curtin Springs that evening, we met a retired winemaker from the Barossa Valley, with a broken front spring on his vehicle. Dave told me to help this chap replace the spring, which required the use of two jacks; one to support the axle, the other to manoeuvre the new spring into position. The old bloke and I wound up spending an hour or so lying in the dust and dark beside his vehicle struggling with the spring, with only the vehicle trouble lamp to see by. We got the broken spring off, but then looked helplessly at each other, as we had no idea how to fit the new one. Fortunately Dave took pity on us and came over and showed us how to finish the job. It was a procedure that would be repeated several times on our own vehicles during the field season.

The winemaker was hoping to find Lasseter's Reef— the fabled 'El Dorado' of central Australia that has puzzled and fascinated everyone who hears of it. In 1930, Harold Bell Lasseter walked into the office of the Australian Workers Union in Sydney with a tale of a rich gold reef he had stumbled upon many years earlier while walking from Alice Springs to Carnarvon in Western Australia. As the Great Depression was then at its worst, Lasseter thought that exploitation of the reef would provide a useful stimulus to the economy. After several meetings, the union leaders set up the Central Australia Gold Exploration Co. Ltd. to go out and relocate the reef under Lasseter's guidance. The expedition met many difficulties and finally gave up, except for Lasseter, who continued on accompanied by a dingo trapper, Johns, who had come into the expedition's base camp west of Alice Springs. Lasseter and Johns eventually fell out, and Lasseter went on alone, only to die of starvation in the Petermann Ranges, after sheltering for several weeks in the cave where the Hull River cuts through the Mannanana Range. In his diary he claimed to have relocated the reef, but did not give its position. Many subsequent expeditions on the ground and from the air have searched to no avail.

The old winemaker intended to go to Lake Christopher, at the western end of the Rawlinson Range in Western Australia, where, he said, 'The spirit of Lasseter will enter into me and tell me where the reef is.' We later learnt that he

The Bloods Range Party north of Mount Olga, Northern Territory, en route to our field area, 1962.

never got to Lake Christopher—an Aboriginal patrol officer had chased him out of the large Aboriginal reserve where Western Australia, South Australia, and the Northern Territory meet, as he had no entry permit.

The following afternoon we arrived at Ayers Rock, now named Uluru, and met Jack Street, the park ranger. He lived in a modern straw hut—a metal frame with straw batts for walls. Jack gave us cold corned beef for lunch, which was absolutely delicious. It would be six weeks before we again ate meat not obtained from a tin.

In those days there was no marked track or hand rail up the Rock, just stone cairns a few centimetres high and many tens of metres apart. That very evening a boy from a school camping tour lost his way in the dusk, slipped down the Rock and fell to his death, in a nearly inaccessible spot well down the side of the Rock but not at its base. A doctor in the rescue party, while they were inching their way to the boy's body, suffered a heart attack—fortunately not fatal—and had to be rescued himself. There is a plaque to the boy's memory at the foot of the Rock, together with several more recent ones.

Jim Kelly and David Forman at lunchtime on the way to Lake Amadeus, 1962.

We went on towards the Olgas, or Kata Tjuta, along the track, long since abandoned, that went in an arc northwest from the Rock and then southwest to the Olgas. We located our fuel dump at a benchmark put in the previous year by a party from the Division of National Mapping, now called the Australian Land and Information Group.

At the fuel dump Jim went to each 44 gallon drum, grabbed it around the top and shook it, exclaiming, 'That's been ratted!' 'And that's been ratted!' as he found them empty. The drums had been sitting there for several weeks, and someone had emptied four or five of them. Fortunately it turned out that enough was left for our needs.

After refuelling we wanted to follow the tracks of the National Mapping party, which had driven north to Lake Amadeus. We couldn't find them, so we had to set off northward at about walking pace to find our own way through irregular sand dunes, desert oaks, and spinifex. Spinifex is a grass (genus *Triodia*) that grows in clumps or rings of nearly rigid spikes with needle sharp tips. Our heavily laden vehicles could not climb the dunes, and so we had to take a circuitous route around each one.

After the first morning, I had been assigned as passenger in the truck. The engine was right inside the cab, with driver and passenger sitting on either side. The screaming of the engine at full revs as it climbed over the low points between dunes was ear shattering. At one stage, the starter motor jammed against the flywheel while we were starting the engine, and made a frightful racket as it tried in vain to turn. A scramble ensued as Jim rushed to disconnect the battery, which was located under the passenger's seat. I had some of my stuff on the floor in front of the seat, which prevented the seat from tipping forward. With much swearing, Jim hurled my things into the spinifex, then went on to deal with the starter motor.

A day or so later, about midday, we reached the southern shore of Lake Amadeus, a vast expanse of salt-encrusted mud, and drove west along the shore. At one place an arm of the lake extends south, which would have added many kilometres to the journey. The channel connecting the arm to the lake was about 50 metres wide, so Dave decided to save time by crossing the channel. All the vehicles crossed successfully except for the overloaded Bedford, which bogged to the axles. We spent the rest of the day digging it clear of the mud, which gave off a sour rotten smell and was so sticky that it stuck to the back of the shovel as much as it did to the front.

Breakfast on the way to Lake Neale, May 1962. Ian Hughes, Jim Kelly, George Kucinskas and Max Mansfield.

Jim Kelly and George Kucinskas inspecting the overloaded Bedford bogged in Lake Amadeus, 1962.

We tried various methods to free the truck. First, we attached the two Landrovers in tandem to tow it out, but the winch cables simply snapped as soon as the load came on, and the truck didn't budge. Next, we made a deadman by burying the truck's spare wheel into the lakeshore after hours of digging. We attached the truck's cable to it and began trying to winch it out. Then we watched in despair as the buried tyre rose out of the ground like a huge mushroom.

The next day, driven to the last resort, we unloaded all eight tons of equipment and carried each piece about 25 metres to the shore. David then drove the truck out unaided. We reloaded it, and set off, having spent some 24 hours at the crossing.

Dave thought we could set up base camp at Mount Unapproachable, a ridge on the northern side of Lake Neale, the next salt lake west of Lake Amadeus. The explorer Ernest Giles in 1872 had hoped to find water there, but his horses bogged continually in the salty mud of the lake, and Giles had to give up the idea. Our overloaded truck found the soft dune sand just as frustrating, and progress was so slow that we too admitted defeat, and camped in spinifex beside a claypan midway between the two lakes.

We had tents for sleeping and eating, and a canvas bucket hanging from a tree for showering, but no toilet facilities, just the instruction to 'Go and do it far away from the camp.' All the food was tinned except for potatoes, onions, and eggs, which were coated with a grease called Kepeg that preserved them for two or, at the most, three weeks. Our staples were tins of Tom Piper braised beefsteak and Imperial sausages—these had no skins, and instead came encased in a solid matrix of fat and jelly, which had to be prised off before cooking. The butter came as 450 gram square blocks in round tins, leaving four D-shaped gaps at the sides. They looked most odd when I opened my first tin, but it tasted excellent, being export quality butter produced by the South Australian Farmers' Union. Max, our cook, baked delicious dampers, and as the weather was cool, these lasted the whole week without going mouldy.

The tinned food was very boring, but after a day's walking on hills, lurching across spinifex, and fixing punctures, we were so hungry that we ate it with gusto. The braised beefsteak, greasy and watery and mixed with gristle, was particularly unappetising, but it was our staple item for dinner, as it had the most meat—80%—of all the tinned meats then available. We improved its taste as

Jim Kelly (left), Max Mansfield, George Kucinskas, and David Forman sinking a deadman at Lake Amadeus, 1962.

best we could with tomato sauce, Worcestershire sauce, pepper, and sometimes by frying it instead of just heating it in the tin. We couldn't face the tinned sausages until all the braised beefsteak was gone. That meant sausages were our only meat for the last five weeks of the field season, and only deep-frying made them palatable.

Powdered milk in those days formed nearly indestructible lumps when water was added, and we became experts at devising ways of making lumpless milk. One good way was to grind together the coffee, cocoa, or Milo with sugar and milk powder before adding the hot water, but the field hands sometimes neglected to do this. Dave complained once or twice, 'Why don't they scratch it up?'

In hot weather towards the end of the field seasons, we missed cool drinks—the vehicles had no fridges. Canvas water bags kept water cool by evaporation through the sides, but we couldn't use them much, as they lost too much water and in any case were ripped to shreds after a few days by the scrub. In my next field season, I devised a contraption which I used from then on. I sewed up a tea towel to form a bag with a draw string, and as soon as we stopped for the night, would wet it, put a tin of fruit or beer in it and hang it up in a tree or put it on the roof of the vehicle. The cotton of the towel slowly stained brown with time and use.

George Kucinskas and damper, Bloods Range, 1962.

Years later, I discovered by accident that BMR had had car fridges from at least the early 1960s, but as the BMR drill parties were stationed next to the BMR store, they had grabbed all the fridges for their own use as soon as they came into the store, leaving none for the geological parties.

Each Monday morning when Dave, myself, Ian and George set off on traverse, Max and Jim would go on a water and mail run; a 300 kilometre, four day round trip to Ayers Rock with a 2,700 litre collapsible rubber water tank in the tray of the truck. Around the fourth and fifth weeks, Max and Jim seemed rather reluctant to embark on their water runs. When we moved base camp about six weeks after, and returned to Ayers Rock, we found that they had been using the same 'shortcut' across the channel at Lake Amadeus where the truck was bogged on the way out. The crossing was now so deep that you couldn't see out of it when you walked down into it. The bottom was littered with the debris of a dozen crossings—logs, branches, lumps of gypsum, even bunches of spinifex.

Before setting out on traverse, we prepared the week's work by examining air photos of the mapping area, selecting the rock exposures we needed to visit, and circling these directly on the air photo with a grease pencil. We also selected and marked the general route for the entire week on the photos so as to avoid impassable areas. We then drove across country and visited the outcrops, where we would spend from half an hour to a few hours, depending on complexity, describing the rocks, measuring structures, deciphering relationships, and sampling if necessary. The geologist almost always walked alone, while the field assistant stayed with the vehicle and fixed punctures. Punctures came in tens per traverse, and two or three at once were common, sometimes four, and on one occasion more than a dozen. The spinifex began at the base camp tent doors, and so we would spend the entire week's traverse lurching at 5 km/h from clump to clump, the only smooth spots being the few metres of loose sand at the top of a sand dune or the needle-covered patch beneath a desert oak.

A five day traverse could range from a few tens to several hundred kilometres, depending on how far the traverse area was situated from the base camp. Washing was restricted to hands and face, and there was no running the tap while brushing our teeth. Towards the end of each week my skin produced a faintly ordurous sweetish smell. We washed the dishes in the water that had heated the tinned food. At dusk each day we'd find the most prickle-free spot and camp for the night—swags under the stars, or rain, and a nice campfire to talk, read or write beside. If we became really bored in the evening, we would set off a bush howitzer—an empty Milo tin with a spoonful of water in it, the lid hammered on tight, a stone placed on the lid, and the whole placed in the campfire to go off when the water boiled. We were off again before 8 a.m. the next morning.

After several weeks on traverse, we began to develop scabby sores on our scalps, and our gums became painful and bled when we brushed our teeth. Weeks-old potatoes and onions were our only 'fresh' food, and we suspected that we were experiencing the early stages of scurvy. We remembered the carton of vitamin C tablets we had seen among the stores we stocked up on in Adelaide, so we ate some and the sores and bleeding gums rapidly cleared up.

The base camp was a welcome relief after a week's traverse. There were tents where we could sleep out of the wind. In the better camps in later years, the tents had canvas floors and wire mesh beds. We could have a brief shower for three nights in a row, which was wonderful after no washing during traverse. It was a great relief also to be able to talk to other people, tell each other the week's stories, and to eat together. The one thing I did regret was that there was seldom a campfire in base camp, so the evenings could be very cold. I often took one of the gas lamps into a vehicle and closed the doors and windows, except for a small air gap, and the heat from the lamp would warm the cabin very nicely. I did this on many a cold evening in the Amadeus Basin, reading, or writing letters, and listening to music on the Traeger radio, totally insulated from the frost outside.

For the six weeks that we operated from the first base camp, we drove northwest along the northern side of Lake Neale, and then headed northeast from there, driving parallel to the sand dunes and visiting low ridges of rock along the way. Except for the smooth going along Lake Neale at the start and end of each traverse, the entire week was dominated by spinifex and desert oaks. Spinifex clumps are so spaced that at least one Landrover wheel is always on top of a clump, so the going is about walking pace, and the vehicle sways like a boat on the ocean.

One of the greatest surprises of my first field season was that we had to use the vehicles as bulldozers to get through scrub where there was no other alternative. Being born and bred in the city, where people treat their cars like expensive furniture and the slightest scratch or dent causes anguish, to be told by the party leader to drive straight into dense mulga scrub in the Petermann Ranges was a revelation. For protection, the Landrovers were fitted with scrub bars, which extended diagonally from the front corners of the vehicle mudguards to the top corners of the windscreen, and then horizontally along the edge of the roof. Steel plates protected the underside of the engine, gearbox, and transfer case.

Over the years I began to feel perturbed by the damage we did to the plants and landscape as we bush-bashed through it. We flattened bushes and small trees, ripped spider webs asunder, and decapitated termite mounds. The pressure of the wheels killed the spinifex, and our wheel tracks could be seen for months or even years afterwards. But I thought to myself that our job had to be done, and the land would eventually recover. In flat areas like the goldfields of Western Australia, cleared tracks bulldozed across an exploration area become the most thickly vegetated parts of the landscape, at times profuse with flowers, as rain-water collects and remains in the tracks long after the surrounding area has dried out.

The Bloods Range area, like much of the Territory, had no public telephones or postal services. We had to communicate with BMR headquarters in Canberra by telegram, using Traeger two-way radios and the Royal Flying Doctor Service—the RFDS—which had set up its own radio communication network in remote Australia many years earlier. Each BMR field party had its own radio call sign, as did every station homestead connected to the RFDS, and when BMR headquarters wanted to tell us something, they sent a telegram.

Several times a day, RFDS would read out the call signs of every telegram they had received, and we listened every lunchtime to the 1 p.m. list. At the conclusion of the list, the RFDS operator would ask, 'Is there any traffic?' All those who had heard their call sign called in, along with anyone else who wanted to send a telegram of their own. The operator wrote down the call signs that he or she had heard, read them out and asked again, 'Is there any further traffic?' This was repeated until all callers were on the list, and then the operator would call up each call sign in turn, read them their telegrams, and/or copy down any telegram the caller wanted to send. By this method, people ordered spare parts and food, made travel arrangements, got medical help, and so on. All the telegrams could be heard by whoever was listening, so everyone knew everyone else's business. Some telegrams, such as those from large cattle stations like Brunette Downs, could be 100 words or more, and if radio conditions were bad, they could take a long time to transmit, with distorted or unintelligible passages repeated until understood.

The most memorable telegram I heard was a 100 word list of vehicle parts sent by Jim the mechanic in my first field season. He carefully said each part name and then a part number, and read the entire list without a break. When George Brown, the Alice Springs RFDS operator and a no nonsense Scot, came on, he was seething. Early on he had become confused as to whether each part number belonged to the preceding or following part name. As Jim had not stopped until he had read the whole interminable list, George, along with much of the rest of the Northern Territory populace, had listened to a largely meaningless monologue that went on and on. When you speak on radio, you cannot hear anyone else talking at the same time. So George had been unable to interrupt Jim and tell him of the problem. When at last he could speak, George well and truly told Jim off: 'That is the worst telegram I've ever heard in my life,' he said in his vigorous Scots accent, and added several other remarks also.

When he'd calmed down, he instructed Jim on how to send such a telegram, still with everyone else listening in: 'This time, say "Stop" after each name of a part and its number, and every now and then ask me if I'm receiving OK.'

A chastened Jim had to repeat the entire telegram, now one third as long again. The whole business was a very useful lesson in telegram procedure.

In addition to communicating with BMR, the other geologists and I also talked to each other on the radios, on a different frequency from the RFDS, at lunchtimes before the telegram session, or sometimes at 6 p.m. if reception was still good, and reported our positions to each other. This was not wholly satisfactory, as we had often moved on from where we said we were by the afternoon or next morning. The procedure of what to do if radio contact was lost was to try again 24 or even 48 hours later, and if that was unsuccessful, send a telegram to BMR headquarters in Canberra and let them solve it. The procedure if the geologist did not return to the vehicle after looking at the geology was for the field hand to go on foot and look for him, and if this was unsuccessful, report the situation at the next lunchtime 'skid'—as the scheduled time for radio communication was known. The radios also had tuneable receiving coils that spanned almost all the AM radio range, and so we were able to keep up with world events at night.

The Landrovers carried two spare wheels, both mounted in the back of the vehicle, vertically above the wheel arches. An extra fuel tank and a water tank, 115 litres each, were fitted one above the other behind the front seats, and so the entire field season was accompanied by the sound of sloshing liquid next to one's ear. The middle of the three front seats had been removed and the space fitted with a plywood box for air photos and various odds and ends.

The two-way radios were mounted on top of the upper tank behind the seats. They used a vertical telescopic aerial that extended to about 11 metres high, and we had to set it up and take it down every time we used the radio. We had a communication skid at lunchtime between the geologists when on traverse, and also set up the aerial on camping for the night.

Bruno Zimmerman and his kitchen on a Landrover tailgate, Davenport Range, 1981.

The aerial had to be inserted into a vertical tube welded to the vehicle; on some vehicles, the tube was mounted at one end of the scrub bars at the front of the vehicle while on others it was mounted below a hole in the vehicle's roof. There was always someone who thought that the natural thing to do was to extend the entire aerial while standing on the ground, and then climb on to the vehicle with the aerial and put it into the holding tube. But as soon as it was fully extended, the 'law of moments' that they had forgotten about came into play and they would stagger around trying to hold up the 11 metre-long swaying aerial, to the shouts and jeers of the spectators. Of course, in the end they had to give up, restore the aerial to its fully telescoped state, climb on to the vehicle and start again.

The gear carried on traverse comprised a tool kit and puncture repair kit; tucker box with eight days' tinned food for two; two swags and personal gear; three billies, a frying pan, cutlery, and enamel plates; sample bags; shovel, axe, and a small medical kit. We carried as little weight as possible, so there were no campstools, table, or lamp for the evenings. In the evenings we sat on the swags, and used the trouble lamp from the tool kit for light—a 21 watt globe in a wire guard that we clamped on to the vehicle battery terminals. The lamps were rather flimsy, and the wires frequently broke off from the clamps, usually at the worst moments. Our kitchen table was the Landrover tailgate. There being no surface water in the area, the 115 litres of water had to last the two of us for five days, both drinking and cooking, so washing, except for hands and teeth, was omitted.

The Landrovers had no synchromesh on 1st or 2nd gear, and the weakness of their springs was legendary. Every field party carried a dozen spare springs as a matter of course, with two, one front and one rear, bolted to the front bumper of each Rover. The unevenness of spinifex terrain is not met in Britain, where the Landrovers originated. By the end of the field season, every Rover was tilted to one side or the other, depending on which of the surviving rear springs was closer to collapse.

The first three weeks in Bloods Range were joint traverses, with geologists, field assistants, and Landrovers all together. I vividly recall the first outcrop we visited—a group of large, yellow boulders of smooth, featureless rock. Here I was, all trained up, and I had no idea what the rock forming my first outcrop was. Dave identified it as silica that had formed by chemical replacement of limestone. By the end of the three weeks, I felt confident enough with the rocks to do my own traverses.

A memorable bog in the salty mud of Lake Neale happened to the lead Landrover on one of these joint traverses. There was nothing to attach the winch cable to, and though digging freed the wheels, they bogged again as soon as they left the cleared stretch.

My Dad had fought in the Middle East during World War II, and recently, after I told him in a letter about the loaded truck's struggle to climb over sand dunes, he had wondered why we weren't carrying sand mats. Recalling his words, I suggested that we fold the four canvas swag covers lengthways and place one in front of each wheel. This worked like a charm.

During this time, the British land speed record holder, Donald Campbell, was building his new 'Bluebird' car, intending to set a new record on Lake Eyre the following year (1963). Our field hands did likewise, comparing notes at the end of each traverse. Their maximum speed on Lake Neale was about 80 km/h, while returning to camp from our last traverse. Campbell's attempts in 1963 were thwarted by rain, and it was not until 1964 that he set a new land speed record of 403.1 miles (648.7 kilometres) per hour on Lake Eyre.

Shortly after making camp one evening northeast of Lake Neale, a thundershower poured down for some twenty minutes, and then cleared. An hour or so later, a large red and yellow 20 cm centipede appeared out of the darkness, crawled unconcernedly through our campsite right past the fire, and disappeared in the opposite direction. This was followed by another, and another, and another. They came every few minutes, two, three, even four at a time for the next 20 minutes or so, and twenty or thirty centipedes must have crawled through the camp, ignoring us completely, and vanished into the night.

The midwinter nights were cold, with temperatures sometimes well below freezing, and if it was windy, the evenings could be pretty uncomfortable. One night near Mount Conner the following year, the wind was so strong that we had to put the billies on the ground downwind of the fire so that the horizontal flames could reach them. We often ate our dinner in the vehicle, and stayed there afterwards, reading or writing and listening to the radio, with the excellent reception that the 11 metre aerial afforded.

On still nights we often woke in the morning with the swags covered in frost and the water in the vehicle tap frozen solid. The taps were made of brass, and if they were forced open, the spindle lost its shape and the tap dripped for evermore. This was not desirable in view of the small supply of water we carried in the vehicle. The thing to do was to fill a billy the night before, but we often forgot.

In our base camp on Angas Downs station the following year, I had the brainwave of warming a slab of sandstone in the kitchen stove after dinner, and then putting it, wrapped, in the bottom of my swag as a bed-warmer. Some years later in the Reynolds Range, northwest of Alice Springs, we would often camp in creek beds to avoid prickles, and one night I had what I thought was an even better idea. I sprinkled hot coals from the campfire on the creek gravel under my stretcher. I returned to the campfire and was just settling down when a cloud of smoke began to rise from the underside of the stretcher. I rushed back and covered the coals with a handful of gravel, after which the swag warmed up nicely.

Around the first week of July, Max and Jim had taken both trucks and gone to Ayers Rock as usual for water, while Dave and I and the field hands went off on traverse. On our return at the end of the week, we found the 1 ton International in a very sorry state. It had overturned on the road between Ayers Rock and Curtin Springs. It was still drivable, but the windscreen was broken, the cabin roof was squashed in, and there was much damage to the other bodywork.

We were all rather subdued as we sat down to dinner, knowing that there would be some sort of accounting for the accident. Dave put his elbows on the table and his hands under his chin, and asked Jim what had happened. Jim maintained that when he and Max arrived at Ayers Rock, there was no mail for us, so he had set off for Curtin Springs to pick up the mail there. As the explanation became more and more involved and complicated, Dave began to doubt Jim's story, and conversation turned into interrogation. Dave never moved, whereas Jim began waving his arms and expostulating. Dave suspected that Jim had been drunk while driving. Jim denied it, but it was clear that Dave did not believe Jim's story. There was no way he could disprove anything that Jim said, and eventually he fell silent, as were we all. The next day it came out that Jim had been driving to Curtin Springs to help at a bus accident, so he was forgiven.

A similar incident occurred towards the end of the field season. During the journey back to Adelaide, several of us noticed that the truck was crabbing sideways. The centre bolt in one of the rear springs must have been partly or wholly sheared through, allowing the rear axle to shift. No one said anything, being very keen to get home rather than delay for repairs. Then, a day away from Adelaide, during lunch at a truck stop in Port Augusta, Dave and Max began arguing about Max's work. Dave was implying that Max was incompetent. In his defence, Max pointed out that Dave had not even noticed the truck's axle was crooked. In anger, Dave told Max, 'You're sacked. You're sacked.' This on the second last day of a four month field season!

Fortunately the others managed to smooth things over, and we reached Adelaide without stopping for repairs.

After six weeks working from our base camp between Lake Amadeus and Lake Neale, we moved to the Petermann Ranges and camped where the Petermann Track crosses the Hull River. Lasseter's Cave, where Lasseter spent the last few weeks of his life, is 17 km to the southeast, in the gap where the Hull River cuts through the Mannanana Range. I mapped the Learmonth Park area, a beautiful pound bordered on three sides by cliffs of quartzite of the Mannanana Range, Ruined Rampart, and Dean Range, and on the fourth by red granite of the Pottoyu Hills, with several large rock holes teeming with shield shrimps.

The explorer Ernest Giles passed through and named it Learmouth Park in 1874 during his second exploration in central Australia, but the name was later changed to Learmonth Park.

The nights in the Petermanns were filled with the howls of dingoes, and it always gave me a thrill of fascination tinged with a hint of fear to hear them. Several times we heard packs of them racing through the rocks and scrub, snarling and biting and ravening. During one traverse near the Ruined Rampart, in the eastern part of Learmonth Park, I was reading in the Rover at about 10 p.m., George, my field hand, being already asleep in his swag. On hearing a sound, I looked up and saw a dingo with his teeth sunk into the foot

TOP: Looking east to Mannanana Range from Learmonth Park, Petermann Ranges, Northern Territory, 1962.

ABOVE: Ruined Rampart from Learmonth Park, Petermann Ranges, 1962.

Shield shrimp, about 3 centimetres long, in a pool on top of Ayers Rock (Uluru), 1963. Shield shrimps, or branchiopods (*Triops australiensis*), are aquatic arthropods with a flattish oval shell with three eyes, a segmented body ending in two spines, and having their gills on their feet. They have remained nearly unchanged since Triassic times, 200 million years ago. They lay their eggs in the mud at the bottom of the pool, and they die when the pool dries out. The eggs hatch whenever the pool fills again.

of George's swag, tugging at it with some vigour. I threw something at the dingo, and it ran off. Many years later I remembered this during the Azaria Chamberlain tragedy at Ayers Rock. On a funnier note, we were camped one night near the Dean Range listening to the song about the pub that ran dry, and as it got to the line about the wild dingoes calling, a dingo started up a howl nearby.

Around September 1962, on traverse near Bloods Range, Ian and I met an Aboriginal family living off the land—they had set fire to a large area of spinifex and were walking along catching small game fleeing from the flames. They had hand-made clothes of hessian, and carried a .22 rifle. The husband spoke some English, and we talked for a while and drew sand maps of the area and where we were going. I much admired that family—they lived wholly off the land, while we were totally dependent on our supplies of food and water in the vehicle.

On another traverse still within Bloods Range, Dave went off to join a BMR geophysical party based near Giles weather station. The party was carrying out a gravity survey by helicopter, and Dave had arranged to use one of the choppers for a day or two to visit remote outcrops in the northwest of our sheet area. In the late afternoon one day that week, Ian and I heard a chopper approaching, and realised it must be our party leader. I felt quite pleased and excited that Dave had gone out of his way to come and see us, naively expecting him to land and say hello. As they flew low over us Dave gesticulated at us, but they flew on without landing. There was indeed no reason to land, and good reason not to, as each landing of a contract helicopter incurs a fee, another fee if you turn the engine off, and so on.

By late September near the end of the field season we had torrential rain for a day and a half, which kept us idle in base camp, because after more than about two centimetres of rain, the ground becomes too boggy to drive across country. So I did not endear myself to our party leader by trying to drive one of the Landrovers across the campsite, and bogging it well and truly. I don't know why I did it. On hearing the roaring of the engine as I tried to get out of the mud, Dave came out of the mess tent and yelled at me through cupped hands. The first part of what he shouted was fortunately inaudible, but it ended with '… stupid idiot'. I managed to dig the vehicle out next day, which was a relief.

By the time we left the area, wildflowers were growing everywhere. On my last traverse in Learmonth Park, it was carpeted wall to wall with flowers and shrubs, birds were singing, and the rockholes were brimful, and teeming with thrashing shield shrimps.

That was the end of my first field season with BMR. We were home in Canberra within a few days of returning our battered vehicles to Adelaide, but Dave went on to write the report on our fieldwork by himself. During the office season he and I produced conflicting interpretations of the large scale structure of the area. He briefly summarised my alternative interpretation in the report, but I was not invited to be co-author. Later work showed that Dave's interpretation was the correct one. I requested and got a transfer to the party that had mapped the adjoining Lake Amadeus sheet, under the leadership of Allan Wells.

3

RAIN, MUD, AND MISHAPS

(1963)

At the beginning of my second season in the Amadeus Basin, in 1963, I joined the Kulgera Party, led by the likeable and kind-hearted Allan Wells. He and I became lifelong friends. The third geologist was a palaeontologist, Stan Skwarko, and the field hands were Bill Woodyatt (mechanic), John Golding and Paul Croker (field assistants), and Jim Morris (cook).

In late May we set off from Adelaide as in the previous year, but this time the trip was wet and muddy. In the evening, north of Coober Pedy, we stopped for dinner by the road, and then Allan decided to drive another 15 kilometres in the dark before making camp. Allan was inspired to do this by our Assistant Chief Geologist's example a few years earlier in the Canning Basin. It was a straight road with no other traffic, so what could go wrong?

Within half an hour, every vehicle in our convoy was in trouble somewhere along the road, and it was here that I found out the treachery of a dirt road after rain. One moment I was driving blithely along, and the next, at much the same speed, smashing blindly through the scrub beside the road. The front wheels had drifted right a touch, and the mud at the edge of the road had grabbed the right wheel and flicked the vehicle up and over the windrow, two wheel trailer and all.

The scrub was just a forest of saplings and shrubs. After a few seconds of hurtling through it, the vehicle stopped, lifted up on top of the mass of smashed vegetation. Steel plates underneath, fitted by the Adelaide store, prevented any damage. Using a nearby tree, I winched the vehicle and trailer back on to the road, turned around and drove back the way I'd come.

TOP: Kulgera Party on the Stuart Highway, en route to field, 1963. Left to right: Allan Wells, Bill Woodyatt, John Golding, Jim Morris and Paul Croker.

ABOVE: Lunchtime on Stuart Highway, 1963, en route to the field. Bill Woodyatt, Paul Croker and Allan Wells.

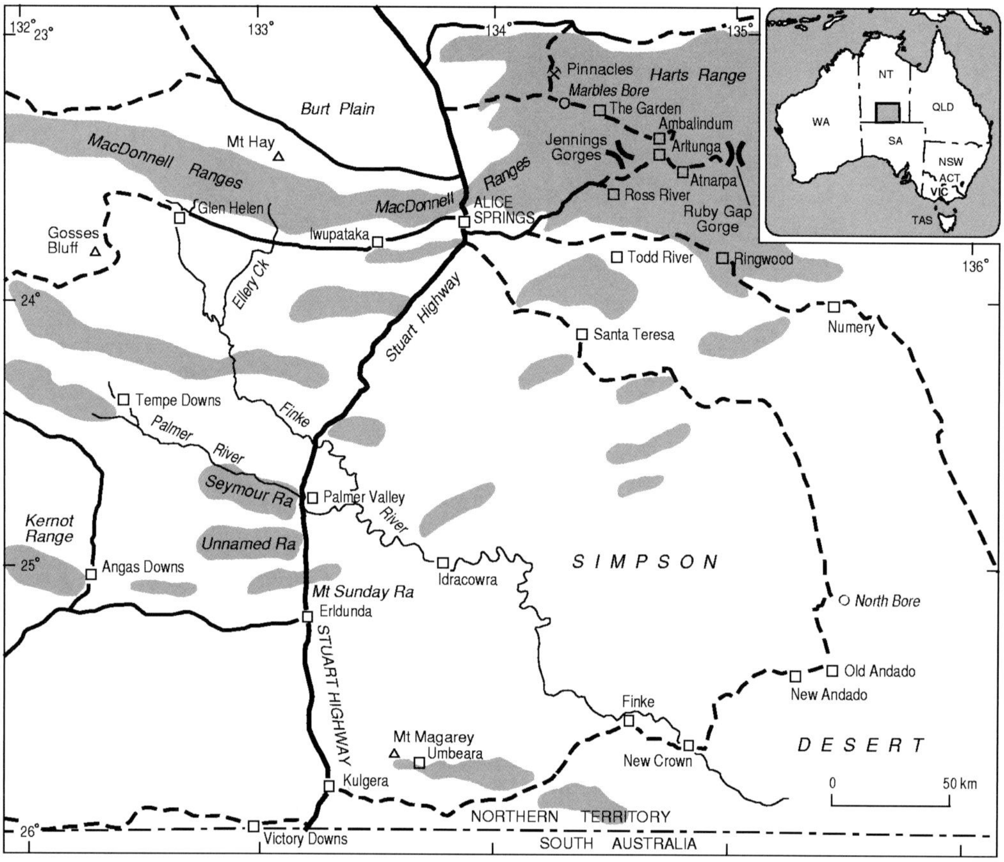

Locality map, south-central Northern Territory.

After driving a few kilometres, I could see headlights in the distance tilted at 45°. These belonged to the truck, which had its left wheels buried completely in the mud at the edge of the road. The other vehicles eventually turned up and we camped off the road beside the truck.

Next morning, we began unloading the bogged truck, and eventually managed to free it. Then we continued on, but we had hadn't gone very far when we came upon Lee Ranford's party who were to map the Henbury sheet, next to ours, with their truck bogged at the side of the road as well. After our practice run earlier in the morning, we were only too glad to help them unload their truck.

The following day around lunchtime, our truck bogged, this time on the other side of a left hand bend. Our mechanic, Bill, decided to try and winch the fully laden truck out backwards. When the tension came on, the bolts holding the front bumper and winch to the truck's chassis sheared off with a bang, and the entire assembly dropped to the ground. So, once again, we unloaded, and then hooked up all the Landrovers like tugboats in line astern, and pulled the truck out quite easily. We loaded the bumper and winch on to the two-wheel trailer behind my vehicle, but they proved too heavy, and the frame of the trailer broke in half a short while later.

We eventually made it to Alice Springs, and had the vehicles repaired at East's Engineering on the north side of town. We then drove out to Angas Downs cattle station southwest of Alice Springs, and set up base camp in mulga and eucalypt scrub near a dam several kilometres from the homestead. This involved considerable cooperation, as erecting tents needs several people, ideally four to hold up the corners on their poles, and a fifth to set up the centre pole, hammer in the tent pegs and attach the guys.

The nights were particularly cold at this camp, sometimes enough to freeze the water in the radiators. The bore water was very hard, and our Persil washing powder had soap and washing soda but no detergent in it. We would boil up a copperful of water, tip in a whole packet of Persil, and watch a gelatinous grey scum many centimetres thick form on top of the water. After skimming this off, we then added another packet of Persil and at last got some usable suds.

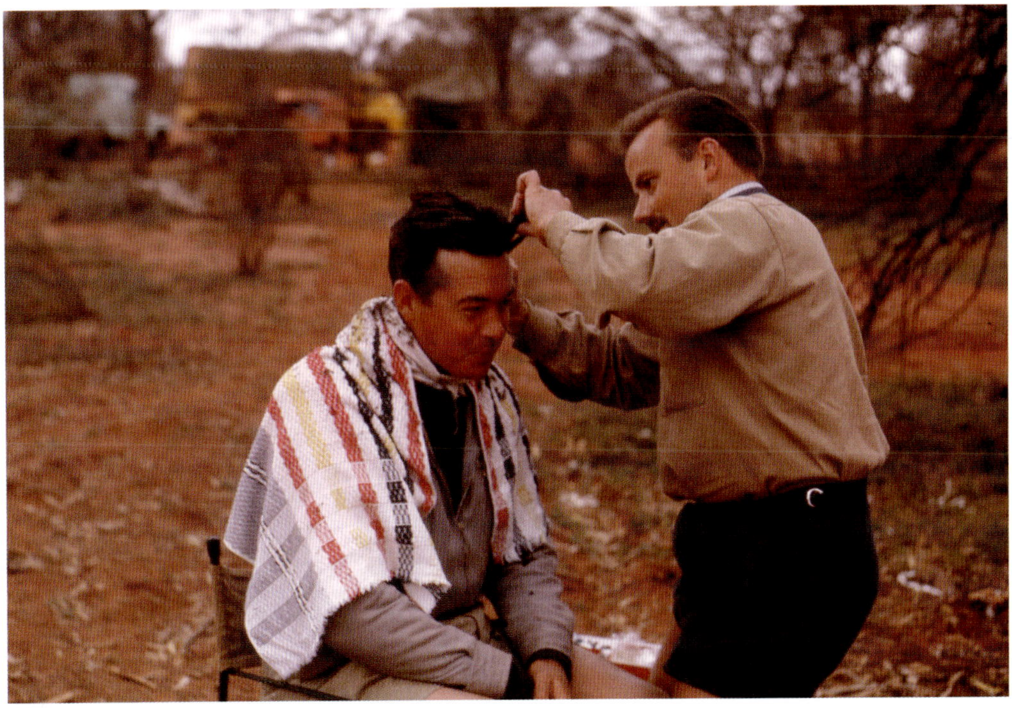

Base camp life, 1963, near Angas Downs; Andy Mikolajczak cuts the hair of party leader Allan Wells.

The first traverse began with all three geologists and the field assistants together, and, by some unspoken agreement, we all took to using the water in Allan's vehicle for cooking and drinking, so as to conserve our own. Nobody mentioned to Allan that it became the custom to go to his vehicle for water, and after three days we separated. Stan and I were in big trouble when we returned to camp at the end of the week—Allan had run out of water soon after we parted, and had to travel many kilometres to Curtin Springs to fill up—with hard and bitter water—and then make the return trip to his mapping area.

It rained again on the weekend after the first traverse, water filling the road near the camp from side to side. When I set out on traverse with John Golding, it was like driving along a canal for several kilometres. On about the second or third day, with the rain gone, we were bumping happily across a grassed-over claypan to an outcrop, when John spied a roughly circular dark patch a few metres across, off to one side. For some reason, I suppose just plain curiosity, he turned and drove straight towards it. I thought he would stop beside it, but he ploughed on and drove right into it.

It turned out to be a deep mass of soft mud in a sort of cauldron in the claypan, and the front wheels went straight down, stopping the vehicle with a jerk. The back wheels fortunately stayed on the hard ground. There were no trees for winching, so we spent the rest of the morning digging. We then had to hack away at the edge of the hole to make a slope to enable the front wheels to climb up and out. Eventually, John got the Rover out backwards by driving it rapidly back and forth and so building up momentum. He had to move the gear lever rapidly from one end to the other of a U-shaped path at the exact moment the vehicle changed direction.

It was a very skilful display, and I thought how much better it would have been to use that driving skill not getting us into the mud hole in the first place. John made up for this lapse when he revealed himself to be a fossil whisperer—he'd wander around on an outcrop and soon find large fossils of nautiloids and other ancient shellfish of great interest to us.

It was on this traverse that I first heard the song 'Cast Your Fate to the Wind', by Vince Guaraldi and Carol Rowe. We had arrived at the last outcrop for the day, a hill of yellow and white sandstone rising out of the mulga scrub. John tossed the swags out on to the ground, raised the aerial, and began setting up camp for the evening. The sun was setting as I walked around on the hill, and I could see for 200 kilometres west to the purple hills and peaks of the Olia Chain and the Petermann Ranges away in the sunset. Suddenly there came this beautiful plaintive tune that captured perfectly the timeless air of the central Australian landscape, with its images of wind, drifting, and emptiness.

Later that season I mapped the western half of the area around Mount Conner, the high flat-topped sandstone hill or mesa rising 300 metres above the plain southeast of Curtin Springs. It was named after Mr M. L. Connor by the explorer William Gosse in 1873. The whole northern side is a vertical cliff above a scree slope, and I spent the last afternoon of the traverse climbing that cliff at the eastern end. Allan had already mapped the mountain itself, but I just wanted to climb the cliff. It looks difficult from a distance, but is actually an easy, if exposed, scramble. The horizontal sandstone has plenty of cracks and ledges for handholds and footholds. The southern side is gullied and an easy, though steep walk, but the vertical northern cliff is more fun. The top shows good examples of 'crazing'—breakage of sandstone into ice cube-like blocks, by the heat of the sun.

I had my worst vehicle accident a few weeks later, in the Seymour Range near the Palmer River, 140 kilometres southwest of Alice Springs. I was supposed to map an unnamed range immediately south of the Seymour Range, and left the Stuart Highway on a bright sunny morning at a point midway between the two ranges. We followed a track going west for a few kilometres till we arrived at a water bore. After that we lurched and heaved over the spinifex. When I got tired of that, I decided to turn north and we headed over to the Seymour Range. I could have just as well gone south to the unnamed range, and to this day I don't really know why I chose the Seymour Range. It just looked nicer and more inviting.

We drove west along a broad valley between the parallel ridges, but the valley gradually became narrower and steeper, and gullies cut into the valley sides became deeper. The gullies had smaller gullies on their sides, and these yet smaller runnels, and soon it became more and more difficult to find a place to cross. Shortly before lunch, the inevitable happened—the vehicle fell over on to its side when the right back wheel slipped into one of the smaller gullies.

Fortunately I had developed a habit of walking beside the vehicle when the drive got too rough. As the driver, however, Bill had not had this luxury. When I looked into the cab, he was lying on the driver's door covered in all my traverse gear—haversack, hammer, notebook, air photos, water bottle, etc.—with a nonplussed expression on his face, but quite unhurt. He climbed out through the passenger's door, and immediately removed the battery from the engine so it would not lose its acid.

Assuming that I would really get it from Allan, I made no mention of the accident at the 12:30 p.m. radio skid. Instead I called up Lee Ranford, the

Somersault in the Seymour Range, 1963. Bill Woodyatt (mechanic) removing the battery.

Henbury Party leader, when he came on the air ten minutes later. He was only a few hours away, and immediately agreed to come over. By the time he and his field hand arrived in the late afternoon, we had the vehicle upright again, using the two jacks in turn and one of the spare wheels as a prop. We had lost some engine oil, but Bill had asked our rescuers to bring some along.

The next morning we took a look at the sandstone ridges on each side of the valley, and then parted. Lee suggested that I not use valley and ridge country for access, but stick to the sand or alluvial flats. Bill and I drove south to the unnamed range, which proved to be lower and with gentler slopes than the Seymour Range, proving that I should have used it in the first place.

In the end I did not get a roasting from Allan at all. He took it as an unlucky mishap, but was puzzled as to why I hadn't called on him first for help.

On another traverse that year, in the Kernot Range, Bill hung up the vehicle by going diagonally into a deep and narrow creek bed. This left two diagonally opposite wheels spinning uselessly in the air, and we spent some time digging away the creek bank under the front bumper until both front wheels touched ground and we got some traction again.

I often went on traverse with Bill, as the geologists changed field hands each week. This was so as not to inflict annoying habits on each other for more than five days at a time. Bill had a habit of setting the campfire only a metre or two from the side of the vehicle, and he liked to build the fire up after dinner. This often caused the petrol to bubble and boil in the fuel tank, which I found terrifying, but we never came to any harm from it. Sadly, Bill died during Cyclone Tracy in Darwin in 1974 while working on a fishing boat.

During that field season, while camped near Angas Downs, groups of us visited Ayers Rock and Mount Olga on our days off, and I went there with Stan and the draftsman, Andy Mikolajczak, both originally from Poland and both normally quite chatty. We climbed the Rock, and found two men already at the top. We said hello and exchanged a few more words, but to my puzzlement Stan and Andy then lapsed into near-complete silence, and I had to make the rest of the conversation as the atmosphere grew noticeably chill. All became clear after the two strangers departed for the climb down, when Stan explained that they were Germans. It was still less than 20 years after the end of World War II.

Meeting on top of Ayers Rock, 1963. Left to right: two German tourists, Andy Mikolajczak, Stan Skwarko.

ABOVE: Looking east from the Olgas (Kata Tjuta) to Ayers Rock (Uluru) 30 km away, with Mount Conner a further 93 kilometres away, barely visible to the right, 1963.

LEFT: The summit of Ayers Rock, 1963. Stan Skwarko, Andy Mikolajczak, and the author.

The first sight of Ayers Rock from a distance—the contrast between the surrounding sea of sand dunes and the steep sides and sheer size of a single hill of completely bare red rock—is extraordinary. The Olgas, on the other hand, are a place of deep shady gullies and canyons with near-vertical sides and a feeling of mystery.

After about two months near Angas Downs, we moved base camp 165 kilometres east to the bank of the Finke River near Idracowra homestead. The place looked most attractive, with stately river gums standing out of what seemed to be firm alluvium. 'I can be very happy here,' I told Allan, and so it became our base for the rest of the season.

The firm alluvium proved to be a thin crust on top of the most fluid bulldust we had ever encountered. It flowed over our feet up to the ankles. Each day as the wind picked up, a pink dust curtain rose with it and covered everything. The sugar on the mess tables turned pink, and looked quite enticing. People appeared out of the pink gloom like phantoms. The vehicles needed low range to drive around the camp, and visibility was sometimes only 20 metres. For years to come, Allan would remind me of my choice of campsite.

Towards the end of the season, Lee Ranford's Henbury Party from Tempe Downs station came down for a visit. A cloud of dust, as always, trailed behind the first vehicle as it drove into our camp, and we expected other vehicles would arrive one by one several minutes apart, each avoiding the dust of its predecessor. To our surprise, the second vehicle was only one or two vehicle lengths behind the first. When its two occupants got out, they looked like dust devils from the desert. They had travelled the whole way like that, as the driver's one quirk was that he couldn't bear to be out of sight of the vehicle in front.

For the barbecue that evening, the station hands from Idracowra produced a gunnysack filled with beef cattle offal. I can still see metre after metre of pink intestine being hauled hand over hand out of the sack. The idea was to cut it into short lengths and grill it. Most of it remained uneaten, and our cook Jim was very puzzled by what to do with the leftovers the next day.

The meal was followed by the traditional evening entertainment at these meetings: running races. At the start signal, a rabble of drunken geologists, field hands and station hands bolted down the course. The enormous cloud of dust lit up by headlights around this mob of running, shouting, and falling men was a memorable sight, missing from daytime athletic meetings in major cities.

Later that same evening, Norma, the wife of one of the Henbury Party's geologists Peter Cook, dropped her wedding ring in the bulldust, where it immediately sank out of sight. We brought the Bedford truck over, turned on its compressor and blew the dust away with an air hose. The search went on for quite a while, in a huge cloud of dust lit up by the truck headlights, accompanied by the the roar of the truck engine, the pounding of the compressor, and the hissing of the air hose, while a crowd of noisy and agitated searchers bent over, peering at the ground through the billowing dust.

Eventually we found Norma's ring, but it was only the next morning, as some of us gathered around the ashes of the campfire, that we discovered Colin Thiele, the mechanic with Lee's party, asleep in his swag by the fireplace, almost completely buried by the bulldust.

A spider encounter happened on traverse that year in the Mount Sunday Range. Our Assistant Chief Geologist John Casey and supervisor Doug Guppy were on a week's visit, and we were all walking up a stony gully on the flank of the range when we came upon a large web spun by a mulga spider. The web hung from a single horizontal strand suspended from one side of the gully to the other, a distance of some five to ten metres. Our field hand immediately grabbed hold of the main strand and tore it apart, sending the web and spider to the bottom of the gully. After we had all gone past the spot, I thought I'd see if a repair was possible, and picked up the two ends of the broken strand. They were as thick as string, not sticky, elastic and surprisingly strong. So I simply tied them together like shoelaces, and the web hung down nearly as well as before, with only a few small rips here and there. We later returned down the same gully, and this time our field hand left the web alone.

Towards the end of the 1963 field season, Bill and I mapped the Kulgera-Umbeara area on the southern margin of the Amadeus Basin. Rabbits being plentiful there, Bill would choose a campsite near a rabbit warren, and set some traps as soon as we stopped for the night. By the time we were ready to start cooking, there was often a rabbit in the trap, sometimes two, and these we stewed up and enjoyed very much, a vast improvement on Tom Piper braised beefsteak. They were also good cold for lunch next day.

One evening during that traverse near Kulgera, at about 8:30, I was reading in the Landrover after dinner by the light of the trouble lamp, when everything—the page, my hands, the inside of the vehicle—suddenly turned green. I looked up, and saw an enormous bright lime green meteor descending to earth directly in front. The head was about the size of a full moon, with an orange aureole around it, which streamed up to the zenith as a red brown turbulent smoky trail. It was completely silent, but appeared so close that I thought it had landed only a few hundred metres away. The next morning I drove off to look for it, without success. The following year, Peter Bloomfield at Atnarpa homestead east of Alice Springs said that he and his family had heard the meteor thunder overhead, and reckoned it had fallen in the northern part of the Simpson Desert, hundreds of kilometres from Kulgera.

Green meteors are quite rare. The green colour results from a trace of copper in iron meteorites, or from magnesium and sodium in stony meteorites, or from doubly ionised oxygen in the excited air around the meteor.

4

THE BEST FIELD SEASON (1964)

The 1964 field season was the last involving the systematic mapping of the Amadeus Basin. Our base camp was in the gorge of the Ross River, about 70 kilometres east of Alice Springs. It was one of the best field seasons I had. Allan Wells, again the party leader, had his wife Sally, daughter Diana, and son Cameron, a few months old, living at the camp. The third geologist, Russell Shaw, was in his first year out of university. The field assistants—Stuart Harper and Jan Rein—were cheerful and willing, and Jim Morris was again our cook. We were camped only a few hundred metres from the Ross River Tourist Chalet, and Gil and Doug Green and their families were very good to us. The campsite was on clean firm river gravel, and the dark red sandstone and yellow dolomite cliffs of the gorge made a beautiful backdrop.

The first traverse, as usual, was a joint one. Lee Ranford and Peter Cook were to map the neighbouring Rodinga map sheet to the south. Allan and I spent a few days with them examining the Amadeus Basin rock sequence along Ellery Creek, west of Alice Springs, and then separated to start the mapping.

I began in the Jay Creek area at the western edge of the Alice Springs sheet, my job being to work eastwards from there to Ross River. On the first night out it rained absolute torrents, and so around 10 p.m. my field hand Jan Rein and I packed up and did the normally forbidden thing of driving across country at night. Fortunately we were only about a kilometre south of the road from Alice to Jay Creek, and had driven in the same way. Nevertheless, steep creek crossings by headlight in pouring rain were very nerve wracking.

By 11:30 or so we reached Alice Springs, went to the old BMR office and mica store[1] and woke up Bluey Kingdom—the resident field hand—and his obese cattle dog. Allan and his field hand were already there, and shortly after, Russell and his field hand also came in, so we were all unexpectedly together, and dry. We went back out next day in sunshine.

Despite the rain in the first week, central Australia was still in drought, and, as in the previous year, pink dust filled the air as soon as the morning breeze came up, coating our sweaty skin, coming into the vehicles through every hole, forming lumps in our nostrils, and adding a little extra discomfort to the hot, dry conditions.

Hale River Party base camp, Ross River Gorge, 1964. The cliff is dolomite laid down in a shallow Early Cambrian sea about 540 million years ago.

Late in the season, on traverse in the Numery area, around September with the weather warming up, we decided to call in at the homestead to tell them we would be working in the area for a while. The wind was getting stronger and stronger and by the time we got there had developed into a sandstorm. We arrived at the front door in a maelstrom of red sand and dust blowing across a hot bare expanse of red dirt surrounding the homestead, with bits of wood and corrugated iron banging and creaking in the searing wind. We knocked at the door several times but there was no answer. We tried the handle, found it unlocked, walked on to a mesh-enclosed veranda coated with red dust on every surface, and through another door into the house. On the other side of yet another door, we found a woman and her young daughter, sitting in the kitchen to escape the dust—and very surprised to see us.

1 Mica was mined in the Harts and Strangways Ranges east of Alice Springs from 1888 to 1960. Its unique electrical and insulating properties and ease of splitting make it useful in the electrical, paint, plastics, cosmetics, and well-drilling industries.

BMR commonly used helicopters for access to remote or inaccessible areas. Our chopper in 1964 had a rather low-powered petrol engine. For the first few days we used it to visit parts of the Alice Springs map area. On the first morning, all three geologists plus pilot climbed on board and set off for the day's work, though the take off seemed laboured and it took a long time to gain altitude.

Flying went all right in the morning, but by the afternoon, with the air warmer and less dense, the helicopter's lack of power became obvious. We started to land near Todd River homestead in a clearing in mulga beside two steep rock walls that met at a right angle, but as we descended the last few metres, the pilot realised that the ground had a slight slope towards the rocks. This meant that we could have landed, but not taken off, as the downdraft from the rotor blades would escape down the slope, and in doing so prevent the chopper from lifting off enough to clear the trees or the rocks. The pilot immediately tried to go back up, but couldn't—the engine did not have enough power to compensate for the downdraft that was already leaking away down slope. We couldn't go anywhere horizontally because of the trees and rocks, so the pilot yelled, 'Allan, hop out.'

Allan did, and very quickly.

Helicopter taking off from base camp, Idracowra station, Finke River, 1963.

Fortunately, the ground was only a metre or so below us. Straight away we were able to rise clear of the mulga and flew several hundred metres to a better landing spot. Allan eventually caught up, unharmed but very surprised. After that, we took only two geologists onboard, plus the pilot.

///

Our cook, Jim Morris, was a good-natured Scot who had been our cook the previous year at Angas Downs and Idracowra, and a field assistant with the Lake Amadeus Party the year before that (1962). He had dark curly hair, laughed and smiled a lot, and had acted in amateur theatricals in Scotland. Whenever there was a song and dance number on the radio from one of the shows, he would dance around the kitchen and mess tent, and he was a good dancer. He made delicious bread, and when he couldn't get baker's yeast he used brewer's yeast, which gave a rather enjoyable beery taste to the bread. Unfortunately he resigned halfway through the field season, as he thought the mechanic, who had also resigned, had been hard done by.

The field assistants took turns as base camp cooks after Jim's departure. Stuart Harper was a cheerful, chatty, good-natured hard worker who laughed a lot, and one Saturday evening he decided to make spaghetti for the whole party. When it was cooked he went outside the mess tent to decant the water, and tilted the pot, which was quite large, on the trunk of a fallen tree. As the angle of tilt increased, the entire mass of spaghetti slipped and fell out on the ground, which was covered in dead gum leaves. He calmly shovelled the well drained spaghetti back into the pot, leaving only a thin layer of spaghetti and leaves on the ground. Unobserved, I saw the whole thing, and said nothing. Stuart was a keen wildlife photographer, and when I met him again some three years later, his photographic equipment filled a suitcase.

He was able to indulge his yen for photography while working as a field hand with Ed Druce. Ed was one of many BMR geologists who used to set up their stretcher and swag pointing out from one of the vehicle wheels, attaching the swag cover to the vehicle to form a wind shelter. He was a palaeontologist and travelled with Stuart into the Georgina Basin collecting fossils.

One day they drove past a dam covered in waterbirds. At daybreak next morning, Stuart jumped into the driver's seat and drove off to the dam with his camera. Ed, still asleep in his swag tethered to the far side of the vehicle, found out the hard way how committed Stuart was to wildlife photography.

The second part of the helicopter survey was spent in the northwest of the Simpson Desert, on the Finke and McDills map sheets, from a fly camp near New Crown homestead. Allan, being the party leader, flew there in the chopper, while the rest of us drove the 500 kilometres from Ross River to New Crown via Alice Springs and Kulgera. We arrived at New Crown near midnight and worn out, the last six hours driving in darkness staring through blowing sand at the nearly invisible track. We were able to locate the fly camp, because they had some lamps and a good campfire going, but we became trapped in what seemed like a maze of wire fences. In the dark, we had no idea where the gates were. We drove for a while along one fence, which was new and strongly tensioned, but got no closer to the campfire. Eventually I had had enough, and decided to take the fence down, drive all the vehicles over it, and put it back up again. We often did this, as it is fairly easy in sandy ground.

Once the fence was down, I stood on the strands, somewhat unsteadily because of the high tension on them, while the vehicles crossed over. The first two crossed without incident, and the third had almost gone over also when suddenly, to my astonishment, I found myself doing a complete somersault in the dark. The tow hook of the vehicle had caught the top strand of the fence and stretched it like a crossbow string until it eventually let go. Fortunately I landed flat on my back with only my ego bruised. That injury was exacerbated by howls of laughter from the field hands, followed by a second dose from the group at the campfire when we reached them a few minutes later. When daylight came, we saw a gate less than a hundred metres from where we crossed.

The return from New Crown to Ross River after the helicopter work was a five day traverse on the Hale River sheet in the northwest of the Simpson Desert. Jim and I drove first to New Andado homestead, and had morning tea with the owners, Mac and Molly Clark. Molly served us fruitcake—it was dark and coarsely textured, and the raisins gleamed on their shiny cut surfaces. I can still taste her fruitcake today.

Mac was killed some years later in an aircraft accident, but Molly stayed on, eventually moving back to the old Andado homestead. She fixed it up, and for years provided accommodation for tourists exploring the Simpson Desert.

After the visit, Jim and I continued on towards Old Andado, and after a few kilometres, came upon a station hand sitting next to his broken-down Landrover beside the track. He thought there was something wrong with the

TOP: Aboriginal stone arrangement, in the northern Simpson Desert, McDills map sheet, 1964. Allan Wells at left.

ABOVE: New Crown homestead, early morning, 1964. Drought, sand, homestead and trees watered from bore with windpump.

Waddy trees (*Acacia peuce*) near North Bore, northwestern Simpson Desert, 1964.

gears. I sent him back in our vehicle to New Andado with Jim, while I sat down to wait beside his vehicle.

When I got bored with waiting, I decided to see if the vehicle really was undrivable. The gear lever felt all right, but the clutch pedal had no resistance—so I thought it was probably the clutch that was the trouble, not the gears. Landrovers in those days had a starter button separate from the ignition switch, and one could if necessary move the vehicle using the starter motor alone. After a few false starts, I found that by starting off in second gear in low range and then getting up to top speed with the starter motor, I could then switch on the ignition without stalling the engine. So off I went back to the homestead with no clutch. I managed to change gears by matching the revolutions in engine and gearbox. I felt pretty smug driving into the homestead yard only a few minutes after Jim and the station hand. We camped that night near North Bore, among the famous waddy trees, acacias that look like a casuarina.

The last traverse was back in the same general area of the Simpson Desert, with Stuart Harper as field hand. It took us a day along dirt roads and station tracks to reach the field area, the last few hours along an old seismic survey track which ran dead straight between the long parallel dunes that typify the Simpson. We camped by the deserted track, and climbed into our swags still covered in grime and sweat.

Daybreak came, bright, clear and still, until the dust arrived when the wind sprang up. After breakfast we discussed how to attack the dunes. Often a small vehicle can cross sand dunes, but each dune needs several attempts, and it uses up a great deal of fuel and time. We had used up a whole day just getting there, and had used more than a quarter of our fuel. Partly deflating the tyres makes dune crossing easy, but the extra width of the tyres can lead to more frequent punctures. Last time with Jim Morris a few weeks earlier, we had left the tyres at their normal pressure, driven parallel to the dunes and gone around their ends. This had greatly slowed the work. In the end we decided that because the drought had withered up so much of the vegetation, we could expect fewer punctures even with deflated tyres. So we let them down to half their normal pressure, and drove straight at the east and steeper face of the first dune. To our surprise and relief the Rover sailed over with perfect ease.

At lunchtime, when we set up the aerial for the usual lunchtime skid, the dust was so thick that a spark jumped across the gap while we were connecting the radio lead to the aerial. All we could hear on the radio was a roar of static, from all the dust in the air. These conditions lasted for the whole three days we were in the dunes. When we finally managed to establish contact on the last day, we found that no one had been the slightest bit worried by our silence.

Although we could cross dunes, much of the traverse still required driving along the valleys, or swales, between the dunes. By the middle of the next day I'd had enough of swaying and lurching over the spinifex in the swales, and decided to drive along the top of a dune, where only a few well separated clumps of spinifex poked out of the clean dry sand. So we drove up to the top of a dune, turned north, and set off along the crest. The crests are irregular and uneven, with peaks and valleys a metre or so in relief. Even so, the going was good, and we weaved our way steadily along the crest. We were beginning to feel quite confident, almost smug, when we came to a single peak forming the entire crest of the dune. The steeper eastern side is where the sand grains are at their angle of rest, and the slightest disturbance causes them to avalanche down the dune face. I assumed Stuart would drive to the left along the gentler western side of the peak, but instead we straddled it, and instantly the Rover tilted over to the right, slid sideways down the dune, and stopped with the driver's door hard against the sand. Stuart was on his side, speechless and still holding the wheel underneath my daypack, hammer, air photographs, and other gear. His expression of puzzlement mixed with worry is with me still. We would have liked to lighten his side of the vehicle to prevent it from completely capsizing, but couldn't. So, while he remained perfectly still, I gingerly opened my door (straight up into the air), climbed out, and scooped the sand from under the wheels on my side. The vehicle gradually tilted back to a more normal angle and Stuart was eventually able to get out. With the two of us digging we got it upright enough to turn and drive down the dune. Much chastened, we resumed our uncomfortable but safer progress along the swale.

With no radio contact, this incident could have ended badly. If the vehicle had rolled, we could have been stuck there until found. Even if we had managed to right it, it would have lost much of its engine oil, as had happened the previous year in the Seymour Range. There were no emergency position-indicating radio beacons (EPIRBs) in those days.

For light relief later in the day, we did a roundup by vehicle of some camels on an alluvial flat, and dug a beautiful yellow-spotted lizard out of its burrow for a photo-shoot.

The afternoon of the last day in the desert was a straight run east across 38 dunes, as counted on the air photographs, to reach an old seismic survey track. It being early October, the heat inside the vehicle was intense. Sand blew into our faces, and we drove for hour after hour into it, up and down the dunes one after the other. We crossed the 38th dune as night was falling, but to our dismay there was no sign of the track. Was it still in front of us, or had we already crossed but not seen it?

A few moments later, to our amazement, a stockman from Andado rode up out of the dusk. Yes, the track was on the other side of the next dune, and there was a new bore a bit farther north up the track too. We thanked our rescuer, drove over the next dune, and there was the track. We drove to the bore hoping for a wash, but to our disappointment the bore was not yet equipped. The water was simply spilling on the ground and had formed a wide sheet of thin red mud. So we drove a few kilometres north and camped in the dark, in the hot, dusty, windy night.

The next morning was bright, sunny, and calm. We had breakfast, packed up, and hopped onboard. But the Rover wouldn't start. There had been no hint of trouble before the previous evening. Suspecting a fuel blockage, we stripped down and cleaned the carburettor and fuel pump and blew out the fuel line, but it still wouldn't go. The reserve fuel tanks had no gauges, so we always used them first. They held more than enough to last an entire normal traverse. We had travelled 160 kilometres from base camp to the traverse area and then crossed dunes for three days. The reserve tank must have run dry at about the moment we switched off the engine the night before. We changed over to the main tank and started with no trouble. We reckoned the tankful would get us back to the base camp. It did, but the next morning when the vehicle was driven away to get water, it went about 50 metres before completely running out of petrol.

5

THESIS WORK

(1966–69)

The Amadeus Basin mapping parties stayed in Canberra in 1965 to write up the previous five years' work as a BMR Bulletin. I left BMR in September for the USA to study for a PhD at Yale, and had a memorable and wonderful three month field season in the Salmon River Mountains of Idaho in the northern summer of 1966. The snow-capped mountain peaks, lakes and torrents of melting snow, and thick green pine forests with squirrels, deer, chipmunks, skunks and bears, were a paradise after the Northern Territory desert. My field hand, Michael Givens, a Yale undergraduate, was a really pleasant chap, and we got on extremely well. We did day traverses out of a national park campground near the small town of McCall, on Payette Lake in western Idaho, and returned there each evening.

Some of the antics of the chipmunks were very endearing—if we gave one a piece of food, it would rush off and take it down its burrow. It being summer, we slept out in the open, and, as time went by one chipmunk became quite bold—I remember one morning being woken by the sound of tiny feet as it scampered across the campground, leapt up on to my rump, then continued off and away. The next morning I was woken by loud and vigorous chattering and twittering. When I opened my eyes I found myself looking straight at a stern and irritated chipmunk clinging to a tree trunk hovering a few centimetres above my head.

TOP: View from Feltham Creek Point (2,743 metres high), Salmon River Mountains, Custer County, Idaho USA, 1966.

ABOVE: Chipmunk as seen on tree trunk, Idaho USA, 1966.

With regard to bears, Mike, being from a city in South Carolina, knew as little as I did. All we knew was that they could smell food from far away, and could be dangerous. While having lunch one day on top of a granite outcrop in thick pinewoods, we heard crashing sounds advancing towards us and realised that a bear was approaching. We immediately and very quietly packed up our sandwiches and sat silent and immobile. After a few more moments, a large brown grizzly appeared, and to our amazement walked unconcernedly past us a few metres below, on the grass at the base of the outcrop.

After we'd resumed breathing, we quietly finished our lunch and carried on mapping for the afternoon. We heard the bear crashing through the undergrowth some distance off an hour or so later, and Mike even called out to it, 'Ho, bear' and so forth, but we did not see it again. Some years later, I heard and read horror stories about bears, including one about a young American geologist who lost both her arms in a bear attack in Alaska in 1977.

/ / /

Idaho is famous for its potatoes—its car number plate slogan for many years was 'Famous Potatoes'—and one could buy postcards showing potato-filled trucks, and stickers saying 'I am an Idaho spud.' Years later I gave one of these stickers to my daughter Alexandra, who stuck it to her primary school suitcase with much glee.

The people in McCall were friendly and hospitable. One afternoon, Mike and I were working along a cutting with beautiful exposures on the main road running west from the town, when a red Ford Mustang cruised up and stopped beside us. The driver, sandy haired and in his forties, introduced himself, asked us what we were doing, and we got into quite a conversation. He said he'd seen us on one or two occasions previously in the cutting, and ended up inviting us to his place to eat that evening. His daughter Louise was there also, having just graduated from school at the start of the summer. We had lunch with her at a café in the town towards the end of the field season. She showed us photos of her high school sweetheart, and they married a few months later. Within a year he was killed in Vietnam.

I also made friends with a girl from Wyoming, Diane Constans, who had a summer job in the McCall hardware store. She dined and took evening swims with us several times in the campground at Payette Lake. We corresponded every year thereafter. Diane is a real back-country girl—goes hunting in the fall, often by herself, bags her one allowed elk, and guts it out at the kill site. She then gets someone to give her a hand to take the carcass off the mountain by backpack or horse pack, on occasion dragging the carcass down by herself to where she can meet them. She then lives off the meat for the winter.

/ / /

Early that summer in Idaho, I realised that I had no holders to hang my notebook and hammer on my belt. I asked around McCall and found a leather and canvas worker who lived with his family just outside the town. His house was a log cabin in a beautiful grassy clearing in the pine forest. I asked him to make some holders for me, and went to his place in the early morning a few days later to pick them up.

I walked across the dewy grass in bright sunshine. When I knocked on the front door, to my consternation his wife opened it, tears streaming down her face. I explained why I had come, and she went back inside. Her mother in law came to the door a few moments later with the new holders, and told me that her son had collapsed and died the day before, soon after finishing them. They were the last things he had made. I was completely stricken, said how sorry I was, paid for the holders, and fled. They turned out to be too small, and I never used them.

The log cabin in the morning sunshine, the huge dark pine trees, and the dewdrops catching the sunlight on the green grass, are some of my most vivid memories of Idaho.

The next year, 1967, I had changed my thesis topic and was back in central Australia studying the squeezed and mangled rocks around Arltunga, some 100 kilometres east of Alice Springs. About 350 million years ago, the rocks in this area were pushed together into a pile of thrust sheets, each sheet being tens or hundreds of square kilometres in extent. My task was to work out the microscopic processes in the rocks that enabled these huge sheets of rock to move and pile up.

Arltunga was central Australia's first town, born out of a gold rush after alluvial gold was discovered in a dry creekbed in 1887. The town was abandoned a few years later when the gold became uneconomic to mine. Some of the old stone buildings were restored about eighty years later, and Arltunga is now a well preserved ghost town. It is part of the 5,000 hectare Arltunga Historical Reserve. In 1964, only one person lived there, at the Crossroads about a kilometre east of the town, and by 1967 no one lived there. My own small base camp was at Arltunga Bore, a few kilometres south of the town. The water in the bore was very salty and unpalatable, so I sometimes got water from a well at the Crossroads. This water was very fresh, and surprisingly warm, given it is only about 10 metres below the surface.

My field hand, Dave Lea, was a storeman from the Department of Supply in Adelaide. As I had been in the USA when BMR arranged my field work, I hadn't been able to interview prospective field hands, so BMR had asked the Department of Supply for a volunteer to travel to Alice Springs and join me there. This meant that I had had to take whoever turned up.

Somewhat to my surprise, Dave did not enjoy walking or climbing, and at the end of his first week he reported that an old shoulder injury had flared up.

Abandoned township of Arltunga, Northern Territory, August 1967; Mount Gordon in background.

I was quite glad to let him return to Adelaide, and I finished the season on my own. He was lucky enough to find two young rock wallabies in a cave while wandering around near the vehicle when I was out on a walk. Of course they'd gone when he showed me the cave on my return.

A few weeks later, while I was walking along a small dry creek bed in hills near Ruby Gap Gorge, a dingo suddenly appeared in the creekbed a short distance in front of me. It trotted off slowly, looking back several times, but eventually disappeared over a hill. After a few more steps I found two dark brown dingo pups on the sand at the entrance to a small cave in the rocky bank of the creek. Their eyes were still closed, and they could just crawl along the sand, but had enough spirit to utter tiny growls when I bent close.

The next year (1968) I took Aimorn, my wife of ten days, out to the Arltunga area for a seven month field season. I had met Aimorn in 1965, during my first year in New Haven, Connecticut, when we were both living in a foreign students' hostel. She was from Thailand and was doing a year of post-doctoral work at the Yale Peabody Museum of Natural History. We corresponded regularly after she returned to Thailand to teach at Chulalongkorn University biology department, and were married in Bangkok in May 1968.

Dingo pup in a creekbed, near Ruby Gap Gorge, Illogwa Creek map sheet, 1967.

There were in fact three marriage ceremonies: an early morning ceremony where nine Buddhist monks came to Aimorn's sister's house at 7 a.m. and chanted for about an hour, finishing with a blessing, followed by breakfast. In the afternoon, there was a ceremony at Chulalongkorn University dining hall where Aimorn's father's law teacher and professor married us at exactly 3:28 p.m. by pouring water on our hands and saying appropriate words, after which everyone else followed suit and wished us luck, etc. Finally, at 9:28 p.m., we sat on the floor in front of Aimorn's mother and father and listened to a homily on making our marriage a success and the rights and responsibilities of each of us. It was in Thai, so I understood nothing of it.

The precise times for the afternoon and night ceremonies were set by the senior monk from the morning ceremony, who used the exact time, date, and day of the week when each of us had been born to choose the times. We then had a week's honeymoon at Caloundra in Queensland, and Aimorn showed me many marine biological wonders in the rock pools there.

We picked up a Landrover, trailer, and stores at Adelaide as usual, and set off north along the then unsealed Stuart Highway. It was raining hard by nightfall of the first day, well north of Port Augusta, and although we had a large tent in the trailer for our intended base camp, we were using only swags for the trip up. They can withstand light rain, but not downpours, so we slept in our swags in the Landrover, myself across the front seats and Aimorn in the back on a shelf that divided the cargo space into upper and lower sections. There is not much room in the back of a short wheelbase Landrover—even less in half of it—but to my amazement, in the morning she successfully rolled up her swag while still inside the vehicle.

We had several more days' rain on the trip to Alice, and this brought out dozens of black hairy caterpillars that crawled about on the wet grass and bushes. Aimorn did not like these, and was very worried that one would crawl into her swag during the night. So every evening she sprayed an unbroken strip of insect repellent right around our campsite to keep them at bay.

By about the third afternoon we were spending more time digging the Landrover out of mud than driving. We pulled off the road to camp near a bulldozer also parked beside the road. We were about to start unpacking when the dozer driver came walking along the road towards us. He had come from a road crew's camp, to put a tin over the dozer's vertical exhaust pipe. After meeting us he offered to tow us through the boggiest part of the road, which the dozer had ripped up just before the rain began. He towed us several kilometres through liquid mud, to a slightly less muddy area past the road camp.

Next morning, we got bogged several times in the first hour or two, so we decided to camp until the road dried out. On the second day there, a man appeared from the north, walking along the road in his socks. He was carrying his boots—'to save them,' he said. He had the clothes he was wearing and the boots, but nothing else. He told us he'd left Granite Downs homestead, a few hundred kilometres north, and was walking to Coober Pedy. We asked if he needed anything.

'Can I have a box of matches, and could you tell the Flying Doctor Service to tell the people at Coober Pedy that I'm on my way there.' That was all, and he continued on south. In the afternoon I followed his footprints in the mud for a few kilometres, and saw that after a while the sock imprints gave way to bare-foot imprints. He was totally mobile in all that mud, while we'd been stuck for days.

That night, quite late, a long wheelbase Landrover came slowly along the road from the north, driven by Ken Collerson, a PhD student from the University of Adelaide returning from fieldwork in the Musgrave Ranges. He said the road was very wet and muddy, 'but you can get through.'

So next morning we set off, making slow but steady progress through the slowly drying mud. Several times cars overtook us in a shower of mud and water, only for us to find them an hour or so later, stuck in a creek crossing till we winched them out. We eventually reached Alice Springs ten days after leaving Adelaide, instead of the usual three or four days.

A day or so later, Aimorn and I set up our base camp at the northern end of Ross River Gorge, across the river from the 1964 campsite. After so much rain, wildflowers grew everywhere and budgies squeaked and chattered in the river gums. We pitched our tent at the base of a red sandstone cliff on a small well grassed sandy flat under a large leaning gum tree, looking out over the river to a dark red sandstone bluff. This was our first house. Beside the tent we set up a tarpaulin stretched around four poles, with a bush shower inside suspended from a branch, and raised and lowered by a rope and pulley. We rolled a large red flagstone from the nearby cliff to make a floor for the shower, and it was still there in 2000 when we again visited the spot. The gum tree was there too, but had collapsed.

After a few days, a BMR drilling party arrived, and we travelled with them to a drill site near Ringwood homestead, 100 kilometres east of Alice Springs. The crew drilled a 260 metre hole into a mass of gypsum, hoping to find potash—any compound of potassium—and perhaps gas. Aimorn and I logged the core and cuttings. The drillers worked two 12 hour shifts per day, and at night, quartz-halogen floodlights revealed an inferno of thick white dust clouds

Our fly camp at the gypsum drilling site on Ringwood station, Northern Territory, 1968, with Aimorn on the left and the mast of the drilling rig rising out of gypsum drilling dust in the right background.

billowing around the clattering drill rig. The rig's air compressor was located to the side of the drilling power plant, so its drive shaft needed two angles; after a week or so of continuous drilling, the universal joint at one of these angles partly melted, and the shaft dropped to the deck. I still have the remains.

Ted Cherry and his crew drilled the hole slowly and carefully, using only the weight of the drill string, and when it was finished the entire 260 metres of drill pipe hung freely in the hole and could be turned by hand. We found neither gas nor potash, but struck a trace of oil, and also the earliest algal fossils, about 825 million years old, in central Australia.

Aimorn worked hard as the drilling was virtually non-stop continuous coring, which produced 5 metre lengths of rock coated with a paste of grease mixed with gypsum dust, all of which we had to wash off by hand with petrol. She became an excellent camp cook, and could turn a tin of corned beef into a really good Thai curry with the spices she had brought along. She taught some of her recipes to Tex, the drill party cook, who managed to replicate her dishes.

While the drilling was in progress, I went with Allan Wells and a group of university teachers on a geological field trip through the Amadeus Basin, while Aimorn stayed in our camp at Ross River. Andrew Glikson, who was mapping at Gosses Bluff west of Alice Springs, was to be the well site geologist while we were away. En route from the drill site, Aimorn and I stopped on the track each to answer a minor call of nature on opposite sides of the vehicle. Andrew's vehicle appeared immediately seemingly from nowhere, and stopped right in front of us. I was sure he'd caught us both with our pants down, but he later told me we were properly clothed when he first spotted us, which was just as well, because he hadn't met either of us before then.

As soon as the university group visited the drill site, several days later, I was to give a short talk about the gypsum body and the drilling results. Earlier, before I left the drill site, the drill had encountered what I thought was dolerite, a rock that begins as hot liquid or magma and would have been injected into the evaporitic gypsum, and then solidified. Andrew had worked in evaporites in Israel before coming to Australia. Fortunately he grabbed me before I could say a word. 'I don't think the rock is dolerite. I'm pretty certain it's anhydrite.' I managed to revise the talk, and sure enough, subsequent work at BMR in Canberra confirmed his field identification.

We worked together several times after that, mainly in the Reynolds Range—Aileron area of the Northern Territory in 1971–75, and then in the Tomkinson Ranges of Western Australia in 1990–91. Our families were good friends for many years, and Andrew and I remain lifelong friends.

After the Ringwood drilling finished, Aimorn and I returned to Ross River Gorge, and I began mapping in the Arltunga area. Aimorn managed to get a job back in New Haven as a research assistant working on sponges in the Yale Peabody Museum. Her job started in September, so I found a field hand, Roger Serong, through the employment office at Melbourne University, and continued fieldwork until mid-January.

Roger had started, then stopped a science degree, joined the Air Force and then left after several months, and was still undecided what to do when the field hand's job came up. He later became an Australian Government architect and private pilot. He was one of the best field assistants I ever had; very cheerful, a willing sample carrier and all day walker in the intense heat of the central Australian summer. We kept in touch every year afterwards.

The fieldwork meant working out in the open at the hottest time of the year, in temperatures often exceeding 40°C in the shade. By midday, standing still to take notes and measurements was a double misery—the heat from the ground hurt our feet, right through our boots and socks, and the flies could settle for long enough to have a good go at our sweaty skin, which left small red and very itchy wounds. In the afternoons, the head of my geological hammer became too hot to touch. We used crispbread biscuits for lunch, as normal bread grew a centimetres-thick crop of green mould in its plastic wrapping after just one day. The dial of my wristwatch turned green with mould, and the watch stopped forever. Worst of all were the snakes. Like us, they found the days too hot, so they came out at night, when we couldn't see them. More than once they came through the tent at our base camp, and through our traverse camp.

The heat working outside in the central Australian summer was so bad that we really found it hard to think. And yet the records kept by explorers a hundred years earlier, such as Charles Sturt and John McDouall Stuart, showed that they carried on through the summers without fuss.

The field season lasted seven and a half months, and became a bit of an ordeal towards the end. I had only employed Roger till Christmas, and so on Christmas Eve he went to join his family in Melbourne, leaving me alone with three weeks' worth of work left to do.

I now had to carry all my rock samples myself, as well as enough water for an all day walk, so the rucksack was quite heavy. One evening on traverse in late December, I was camped in a narrow sandy creek bed near The Garden homestead, sitting on my swag after dinner gazing into the fire, when a 2

Roger Serong inside a flooded gold mine, White Range, Arltunga area, 1968.

metre brown snake came slithering along the creek bed straight towards me. I frantically climbed up on to the roof of the Landrover, while the snake, no doubt just as surprised, climbed up the bank of the creek and continued past and off into the night. From then on I slept on the Landrover roof.

A week or so later, alone and missing Aimorn, I remember stopping the vehicle at an outcrop near Arltunga, walking up to the rocks and seeing so many different rock types that my brain went completely blank. To my surprise I then burst into tears and went and sat in the vehicle. After recovering I turned around and drove back to my base camp in Ross River Gorge, gave myself the next day off and lay around reading a novel for the entire day. I went back out again the following day, greatly refreshed, and finished the fieldwork.

As I was packing up to leave the field in mid-January 1969, I received a telegram from BMR headquarters asking me to meet three geologists from overseas who were attending a conference on meteorites in Canberra, and show them Gosses Bluff, near Hermannsburg. Gosses Bluff is a large exposed impact structure that formed about 140 million years ago when a comet or meteorite struck the Earth. It was mapped jointly by Andrew Glikson of BMR and two geologists of the United States Geological Survey in 1968 as part of the United States' Apollo Program to put men on the Moon. The reason for mapping a terrestrial impact structure was to give the astronauts some idea of what to expect when they landed on the Moon, which has thousands of impact craters, from large to small. During the conference, however, a large volcano in Indonesia began erupting, and two of the scheduled three visitors went there instead. I met only one, Tom Ahrens from the California Institute of Technology, at the Alice Springs airport, and took him out to Gosses Bluff.

The next morning, at 8 a.m., as we were about to climb to the top of the bluff, it was already fiercely hot. I told Tom to carry water, but he took a long drink there and then and said that would be enough for the climb up and back. When we reached the top he drank all my water. We were both very thirsty by the time we got down again. However, he was a good companion on the trip, a fount of knowledge on impact structures, and later he sent me some nice colour photos of the shattered rocks.

That, however, was not quite the end of the field season. I flew to Adelaide and while waiting in the Adelaide airport to change planes for Melbourne, I heard an announcement asking me to go to the information counter. Very puzzled, I went to the counter and was handed a telegram which said 'Stewart to return to Alice Springs.' It added that I was to await the arrival there of some company geologists, and take them on an inspection tour of the Amadeus Basin. My puzzlement turned to amazement when I read the signature line. The company—Cundill Meyers and Associates—had issued the telegram, not the BMR. After seven and a half months of fieldwork, the last four in the summer, I was in no mood to do any more, especially when ordered by someone masquerading as my employer. A few hours later I sent the company a telegram from Melbourne telling them that I had left the field.

When I called in at BMR in Canberra on my way to the USA, I found out the whole story. The company had talked to the acting chief of BMR's Sedimentary Section during January, to see if someone could take their geologists on an inspection tour of the Amadeus Basin, and been told that I was the only BMR geologist out there, and if they wished, they could ask me if I would take them on a tour. They had taken it upon themselves to order me back to Alice Springs and wait for them.

I never heard from them again, not even on my return to Canberra, and presumably they never made it out to the Red Centre.

6

FIRST TIME AS PARTY LEADER

(1971–72)

I went back to BMR in January 1971, and joined the Arunta Project led by Russell Shaw, my friend and tent mate from the 1964 field season. We set up base camp at Marbles Bore northeast of Alice Springs, a few hundred metres off the main road heading east from the Stuart Highway to The Garden and Ambalindum homesteads. I was party leader for the first two months until Russell arrived.

It was a very large party. The field season lasted from mid-May to mid-October, but each geologist was there for only two to three months. The party's job was to map in detail the igneous and metamorphic rocks of the Alice Springs and Napperby map sheets, which included numerous small abandoned mines, most of them copper or gold mines, but no large deposits. The other geologists were Gladys Warren, a former school teacher turned geologist, with an astounding memory, very kind hearted and sensitive, Andrew Glikson, Mike Rickard, from the Geology Department at the Australian National University, Mike's graduate students John Funk and Roger Marjoribanks, and Mohammed Khan, a graduate student from Monash University.

We had five field assistants, two mechanics, and a cook. One of the mechanics, Neville Cole, brought his own caravan and near-term expectant wife to the camp. She went to Alice Springs for the birth, then returned to the camp with the baby and lived with us until the end of the field season.

Partway through the season, we inherited two more graduate students, Peter Woodford and Alastair Allen, as well as S. Iyer, a post-doctoral fellow from India, all from the University of Queensland. They came to us when they suddenly lost funding that had until then been provided by a mining company. Roger Marjoribanks worked in the neighbouring Hermannsburg map sheet to the west, on a similar problem to my own thesis work around Arltunga in 1967–68.

For most of the field season we used black and white air photos; Russell had ordered colour photos, but they arrived only for the last month of fieldwork. So before leaving for the field we enlarged the black and white air photos by two, giving them a scale of about 1:23 000. This made the photos very large, nearly half a metre on each side, and rather unwieldy to carry through the scrub. So I cut mine into quarters. What we hadn't realised when we enlarged them was that we could no longer use them as stereoscopic pairs. Our heads would have needed to be enlarged by two also to make our eyes far enough apart. So we plotted our observation points on the enlarged black and white photos, copied every point to its corresponding position on the colour photos when they arrived later, and then did the stereoscopic photo-interpretation. The colour photos were magnificent for distinguishing different rock units after years of struggling with subtle shades of grey on black and white photos.

The Marbles Bore base camp was in a large clearing in open scrub, and after several rainy seasons in the late 1960s, the vegetation had produced masses of seeds. These had since dried out, and lay on the ground as an invisible carpet of thorns and prickles.

Coping with thorns and prickles is a big part of living in the outback. They catch in your socks, and work their way down between boots and socks. The most difficult part is taking your trousers and underpants off. If you leave your boots on, prickles leap from your boots on to the inside of these garments, and some are tiny and hard to see, let alone remove, from clothes or skin. On his first morning in camp, Iyer ran barefoot across the camp ground towards the shower, but suddenly stopped halfway and danced on the spot to remove the prickles first from one foot and then the other, trying at the same time not to stand on more prickles with the foot just cleaned.

The party was so big that at times things got a little out of control. The cook left partway through the season, and in August three field hands overstayed a weekend's leave to attend the Harts Range races, and then damaged one of the vehicles on the way back.

The next morning I went out on traverse with one of the three, along the road where the accident had happened, so I asked him to show me the place. It was an amazing sight—the wheel tracks showed that the vehicle had veered off the road on to a smooth gravel flat, then done several circles at speed on full front wheel lock, spraying a shower of dirt outwards. It had then headed straight for a ditch dug obliquely to the road by a grader to allow water to drain off. The dirt from these ditches is left as a ridge along one side of the ditch, and the vehicle had hit the ridge, flown through the air for several metres, and landed on the far side of the ditch. The reason for the damage to the front axle was now clear. My field hand explained that the driver had suddenly gone crazy, done the wheelies for fun, and not seen the drainage ditch until it was too late. Russell sacked the three of them, and the Department of Supply sent replacements from Adelaide.

Rodney Richardson, one of the replacements, was a loner from the Queensland bush. Towards the end of the field season, he and I did a 12 day traverse in the Reynolds Range area, and I found him to be good natured and willing. He told me he had once driven a flock of sheep along the road from the Northern Territory to Queensland while driving a utility truck. I immediately had visions of wayward sheep running away from the roadside and being very hard to round up using only a vehicle, so I asked if he'd had anyone to help. 'Always on me own,' he'd drawl, 'Always on me own.'

During the Reynolds Range traverse we did an all day walk to the summit of Mount Thomas and back. I was a bit surprised when we got back to the vehicle at dusk to find that Rod was so exhausted he could eat nothing.

The first cook at Marbles Bore was a very pleasant Englishman, Steve Coupland, who had cooked on oilrigs in the North Sea. Our camp was about three hours' drive from Alice Springs, and so well within reach for supply runs, and had a large kitchen caravan with a refrigerator, freezer, and restaurant-sized gas stove. Steve produced some of the most delicious meals I'd ever had in a camp, and all of them different. He enjoyed lots of company, and more than once two British nurses, who lived at Ambalindum homestead, about 50 kilometres east of our campsite, to care for the elderly station owners Bill and Pauline Cavenagh, came down to our camp after dinner for music and dancing in the kitchen caravan. Tom Jones was their favourite singer.

On one occasion, I was first into the mess tent for breakfast and found a total stranger asleep on the kitchen caravan floor—Steve had met him on an Alice Springs supply run and brought him back to camp.

The good times ended when Russell went through three months' mess accounts with Steve, and found that we were seriously in debt to suppliers in Alice Springs. Steve had been cooking to oil exploration company standards, serving large, high quality meals and throwing out the leftovers. This was too lavish for our field party, and after Russell asked him to cut down, he departed, very miffed.

His replacement was Len Griffin, a slightly built, talkative, and very funny South Australian who had worked for Russell a year or so earlier. Len was a wizard at economical camp cooking and using leftovers, and had the accounts back in the black by the end of the field season.

/ / /

We often used to call in on Bill and Pauline Cavenagh at Ambalindum homestead, preferably at morning or afternoon teatime. We would look at Bill's large and excellent rock and mineral collection, which included a hollow geode or thunder egg from Mexico, which contained liquid—liquid that was millions of years old—that you could hear splashing around inside when you shook it.

Bill Cavenagh's son Don and his wife Neva lived a few kilometres east of Ambalindum at Claraville homestead, with their young son, a toddler. One evening, the little boy was exploring the floor of a station vehicle when he found a bottle under the driver's seat and licked its top. Within a few minutes he collapsed and died—the bottle had contained strychnine for baiting dingoes. Bill had put the bottle under the car seat. To make things worse, the Royal Flying Doctor Service alarm system, which at the time required alternate blasts on two whistles of different frequency over the radio to activate an alarm in the RFDS base in Alice Springs, wasn't working. When medical help eventually did come, it was far too late.

The RFDS alarm system failed a second time after an event that occurred one night in base camp when we were all asleep. At around 3 a.m., I woke to hear a distant car engine revving at high speed and slowly drawing nearer. After a while headlights appeared on the track to our camp, so I got up to meet whoever was coming. Eventually a Landrover drove slowly into camp with the engine screaming at maximum revs in low range, and with a list to one side. The driver turned out to be a miner from a small copper mine at the Pinnacles some 15 kilometres north of our camp. He and his wife had been returning to the mine late at night from Alice Springs, where he had bought her a sewing machine, when they ran off the main road at a sharp bend. She was thrown out of the vehicle and badly injured her back. He had managed to lift her into the vehicle, but discovered that the tyre and tube had been completely torn off from one of the front wheels. He had driven 25 kilometres to our camp on the wheel rim.

We set up a camp bed in the tray of our truck, put the patient, who was in great pain, on it, and set off slowly towards Alice Springs. Once again, the whistle-activated radio alarm did not work, and it wasn't until nearly dawn that I managed to get in touch with the RFDS. They sent an ambulance to meet the truck, but it was almost in Alice when they met up. Fortunately this primitive alarm system was eventually replaced by satellite telephones.

The following year, 1972, I took my own regional mapping party to the Reynolds Range area, about 100 kilometres northwest of Alice Springs, and set up base camp a few kilometres from Pine Hill homestead. The area is a complex of igneous and metamorphic rocks, with a few small abandoned mines that had been worked for tin and copper early in the 20th century. The other geologists were Lyall Offe, a tall young recent graduate from Adelaide University, talkative, keen to learn, and utterly trustworthy, and Gladys Warren. It was an eventful field season—the cook Bill McGrath, Rodney Richardson, the field hand who preferred his own company, and the mechanic Georg Schmolke left at different times, leaving us with a solitary field hand, Ian Harverson, for the last few weeks. We also had a draftswoman for the last half of the season, Di Pillinger, who drove down from the BMR office in Darwin and added much sparkle to the camp.

That was the first year we used short wheelbase Toyota Landcruisers. Ian's vehicle soon became very difficult to start. Bill the cook suggested pouring a spoonful of petrol into the carburettor before starting for the first time each day, and this usually worked. On one occasion in the base camp, late in the evening I went to Ian's vehicle to retrieve something. It was parked beside the field hands' tent, and while I was rummaging around, a voice out of the darkness said, 'A stubby says you can't start it.'

I immediately hopped in, pumped the accelerator pedal for what seemed like ages, pressed the starter, and to my delight she started first go. I drank Ian's stubby with great satisfaction.

Ian had come out from England only a few years earlier, and was fascinated by the budgerigars, which were very abundant that year, and he spent much time photographing them. He was much taken by their habit of nesting in holes in

TOP: 1972 base camp on Pine Hill station, Reynolds Range area, Northern Territory. Kitchen/mess tent on the right, drafting caravan with sample drums and Sue Schmolke in centre, washing machine and clothesline (with nappies) on the left.

ABOVE: Lyall Offe and Di Pillinger, last evening in base camp, Reynolds Range area, 1972.

trees, and tried several times, unsuccessfully, to photograph one at the entrance to its hole. Luckily one day we found a pair of budgies nesting in a small gum tree right where we had stopped for lunch in a narrow sandy creek. The nest was at the bottom of a long hollow dead branch, which had several holes along its side where smaller branches and twigs had fallen off. While we were having lunch, we could glimpse one or other of the budgies scuttling up and down the inside of the branch on their way to or from the nest. Ian made several attempts to snap a budgie at the entrance to the branch, but each time it flew off too quickly. Eventually, Ian got me to stand in the creek bed on the far side of the branch, to keep track of the next budgie on its way up and tell him how far it was along, so he would be ready to take the photo at the very instant the budgie appeared at the entrance. Ian got his photo, only to find when it was developed that it also included my own blurred face with mouth agape, yelling, in the background.

We had a week of helicopter work on the Mount Theo and Mount Peake 1:250 000 Sheets, and set up a fly camp near the ruins of the old Mount Doreen homestead. On the first evening, we met an anthropologist from the Australian National University, Nick Petersen, who was doing field work with the Warlpiri Aboriginals there. After we'd said hello, Nick asked us, as a favour, if we could take one of the Warlpiri men with us each day in the helicopter, and record the Aboriginal names of topographic features we flew over or visited, as his vehicle did not have the range to reach the most distant parts of his area. It would mean having only one geologist in the helicopter instead of two, but in the end we agreed, and recorded many names and their locations on air photos and in due course on our geological maps. Later on I sent Di and Ian out on a week's traverse, and they collected many more names from the Napperby and Yuendumu areas. Back in Canberra, we submitted the names to the Northern Territory Place Names Committee, and many, such as Anmatjira Range and Ngalurbindi Hills, now appear on maps of central Australia.

The road trip back along the unsealed Tanami Road from the fly camp to our base camp was interrupted by several accidents. Di and I flew back in the chopper, and expected the others to meet us there after driving down the Tanami Road, then taking a road through Napperby so they could get to the base camp via Aileron, on the bitumen Stuart Highway. They still hadn't arrived when night fell, and there was no sign of them until around midnight, when Lyall and the helicopter mechanic finally drove in, exhausted, dirty, near breaking point, and without Gladys.

Emu Dreaming and a Warlpiri elder at Nungurmanu, Mount Theo map sheet, 1972. 'Dreaming' refers to stories and songs of the Ancestors, which serve as a mental map of places where food and water can be found. The stones at this place outline an emu's footprint, with the central toe omitted. A rockhole about 1.5 metres deep (with a dead emu at the bottom when we visited) is located about 2.5 metres left of and outside the picture.

Around dusk, Lyall, in the truck and towing the water trailer, and Gladys and the helicopter mechanic, driving a Landcruiser each, had gone different ways at the intersection 22 kilometres northwest of Tilmouth Crossing. By the time Lyall had turned back to look for them, Gladys and the mechanic had continued down the Tanami Road to the Crossing, where they turned left and headed north towards Napperby homestead. At one stage, Lyall tried to do a U-turn but was prevented by the sloping banks at the roadsides. He had to uncouple the trailer, which instantly tilted forward and jammed his hand between the coupling and the ground. Luckily the ground was sandy enough to give, and he wasn't badly injured.

He only caught up to them between Napperby and Aileron. He found them stopped beside a damaged Landcruiser. The helicopter mechanic had run off a sharp bend and crashed into a tree. He wasn't hurt except for a scratch on his forehead, but the damaged vehicle was undrivable. Gladys volunteered to camp beside it for the night, while Lyall and the mechanic continued on to our base camp.

The next day Lyall and I retrieved the damaged vehicle and slowly towed it to Aileron. The front wheels were splayed out at a wide angle, but could still rotate. It took us another two days to tow it 80 kilometres along the gravel verge of the Stuart Highway to Alice Springs for repair, scrubbing virtually all the rubber off the front tyres in the process. For the last 15 kilometres into Alice, we could not use the verge, and had to drag it along the bitumen. We probably set the record for the longest skid marks ever.

A happier event occurred at the Aileron Hotel. Lyall and I had gone to Aileron to pick up the mail after a traverse, and while having a beer in the bar, noticed a trapdoor in the floor.

'What's down there?' we asked the barman.

'I don't know,' he said. 'I've only been here a few months, and never gone down there.'

We persuaded him to open the trapdoor, and went down into an underground cellar. Lying in boxes of wood shavings were several dozen bottles of champagne. They must have been there for years; some had turned brown,

Di Pillinger and helicopter on granite boulder, Anmatjira Range, Northern Territory, 1972.

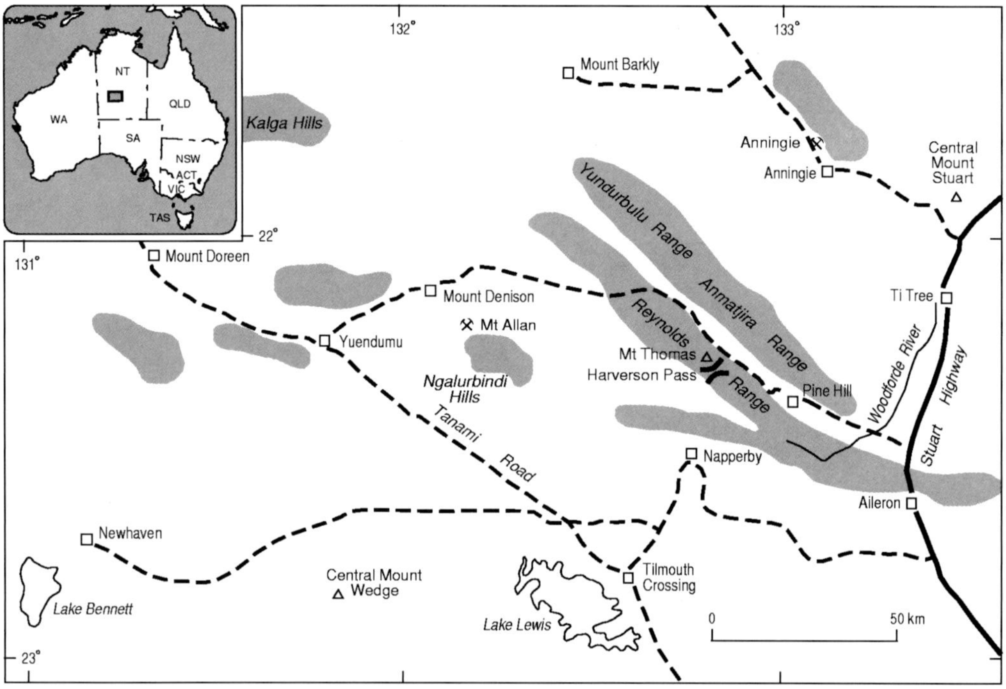

Locality map, Reynolds Range—Aileron area, Northern Territory.

and some bottles had leaked, though the corks were still wired on. We were able to convince the barman that nothing could save the rest and that they would soon go the way of those already spoilt, but as a favour we would take them off his hands for $1.50 a bottle. After some hesitation he agreed, and we had champagne dinners for the rest of the field season.

Mike Rickard from the Australian National University worked from our camp for the first few weeks of that season. He had an endless supply of fascinating and humorous stories from Ireland, Canada, and Fiji. Our rather clever hot water system for the camp shower consisted of a 200-litre drum of cold water up a tree, and a similar drum heated by a fire at ground level. The upper drum provided the pressure to push hot water from the lower drum up into the showerhead. The upper drum had to be refilled with a pump from the main rubber water tank mounted on our four wheel trailer. One cold and windy evening after we had all returned from traverse, I was in the drafting caravan near midnight when a voice sounded from the doorway, with a polite enquiry. I turned and found Mikes' large, bearded, untanned figure in the doorway, shivering and clad only in a few soapsuds. No one had told him how to refill the upper drum.

Out on traverse, it was an unwritten rule not to wash except hands and teeth, because we carried only a small amount of water. By the end of a day's walking and climbing, often interspersed with jacking up a vehicle, changing a wheel, and on occasions digging the vehicle out of sand or mud, we got pretty dirty. Afterwards your skin feels rough and stiff, and it grips and then releases your shirtsleeves and trousers instead of letting them slide freely when you move around. Sleeping unwashed in a swag is even worse—it clings to almost every part of you, and smells terrible after a week.

During that field season, I decided to find out if there was, after all, enough water in our vehicle tanks to allow a mugful each evening for a wash. It turned out there was, and from then on I washed every evening, a fairly chilly ordeal on windy mid-winter evenings. It was even worse if I had to stop for the evening skid with the party leader and carry on a radio conversation while standing wet and shivering beside the vehicle. Being able to spend the rest of the evening with a clean skin made the initial chill worthwhile.

After a year or two, I worked out how to rig up a windbreak for my evening bath. We would park the vehicle so that the passenger's door was downwind, prop the vehicle door open, and stand my assembled stretcher vertically on end against the outside of the door. Then I'd jam my swag, still rolled up, in the space between the vehicle and the ground under the passenger's doorway. Finally, I'd pack my fake fur-lined Mighty Mac in the space between the windscreen and the open door, completing a small but effective shelter from the cold desert wind.

That year, 1972, I used to ring up Aimorn in Canberra from the telephone in the Aileron hotel. The phone was located behind the bar and right beside the refrigeration motor, which frequently switched on and off during a conversation, so we had to scream down the phone to each other, and even then could barely hear. I found it extremely frustrating, especially when a problem arose with a young female boarder we had taken into our house to be company for Aimorn who was new to Canberra. The boarder began to be pestered by her former partner, who took to coming to our house at night and shouting and banging on the front door. Trying to sort this out by yelling into a telephone beside a deafening industrial refrigerator was nigh on impossible. Fortunately Mum took on her role as the Colonel's wife, flew up to Canberra from Melbourne and frightened him off with her Army invective. The girl realised that she was causing Aimorn and our two year-old daughter Alexandra a great deal of trouble and fright, and subsequently moved out.

In the second week of the field season, an ingrowing hair on Rodney Richardson's left cheek became infected, and he had to stay in Alice Springs for hospital treatment for five days. When he left the hospital, he was told to report back twice a day for the next ten days or so for the dressing to be changed. I was flabbergasted, as this would seriously disrupt our work.

I convinced the doctor that we could dress the wound ourselves if he would show me exactly what to do. Every morning and evening, in base camp or on traverse, we boiled up forceps and scissors in a frying pan of water on the fire, swabbed the wound three times with disinfectant, cut a piece of paraffin gauze and packed it in and over the wound, and then covered that with gauze and sticking plaster. The wound healed well, but left a small vertical slit in Rod's cheek, with two tiny lips of skin on each side, and it annoyed him considerably.

On that same traverse Lance Black, a geochronologist newly recruited to BMR, joined us for several weeks, and we collected many rock samples to determine when the rocks had formed. This process needs large pieces of fresh rock, which we obtained by drilling several holes into a rock face with a portable drill, inserting gelignite, setting and lighting black powder fuses with detonators, and blasting the rock apart. Rod was an experienced shot firer, and we obtained many useful samples in the Reynolds Range area.

Late one afternoon, however, part way up the flank of the Yundurbulu Range, the blasting had been only partly successful. The deeper parts of the drill holes were still intact, and we could see unexploded gelignite in them. Unexploded gelignite cannot be left, in case it spontaneously blows up later. The only way to destroy it is to set off more explosive next to it. As it was then dusk, we decided to leave things until morning, and made camp on the alluvial flat at the foot of the range.

The next morning after breakfast, as we were all climbing back to the site of the unexploded gelignite, we heard two deafening explosions right beside us. We were momentarily terrified, thinking the unexploded gelignite had blown up of its own accord, but there was no cloud of dust nor shower of rock fragments. The Concorde aircraft had just flown over us on its inaugural visit to Australia, and the explosions had been its sonic boom, caused by flying faster than sound.

We'd heard that it was due in Sydney that morning, and soon realised what had happened. After we calmed down, we resumed the climb, carefully laid fresh sticks of gelignite beside the holes, and blew up the lot, this time successfully. From then on we called this sample site the 'Concorde locality'.

Rod went on to do several weeks' work with Lance in the Alice Springs map area to collect more geochronology samples, working out of Russell Shaw's base camp. In August I returned to Canberra for three weeks, and during that time Rod and Lance fell out and Rod resigned. I never saw him again.

Ian Harverson, in my party, and Peter Jennings, a field assistant in Russell's party, were experienced mountain climbers who had climbed in New Zealand. They spent much of their spare time planning and organising a climbing expedition to the Colombian Andes for the following year. They spent one weekend climbing the precipitous southeast face of Mount Thomas, the highest peak in the Reynolds Range, and on another occasion I watched in amazement as Ian climbed up an apparently smooth and featureless granite boulder several metres high. He spread his arms and legs in all directions and went up like a spider, using the edges of exfoliation weathering sheets only a few millimetres thick to cling to the side of the boulder.

Near the end of the field season, our boss in Canberra, John Casey, realised that a BMR drilling party would be travelling through the Mount Theo map area along the Tanami Road, and that this would be a good chance to drill holes beside the road to get some subsurface information. I asked Lyall to meet the drillers and log the holes—to describe the chips of rock brought up by the drilling, describe core samples, and measure rock properties such as electrical resistivity, density and radioactivity with sensors lowered down the hole. Inevitably the drilling took longer than expected, so Lyall was delayed with the drilling party, until after Ian and I had packed up the camp and departed for Adelaide.

When we met up in Canberra a week or two later, Lyall was a bit scornful of the logging arrangements. He had arranged to pick up a suitcase-sized electronic logger, called a Portalogger, from another field party as they travelled to Alice Springs from the Tanami area, far to our northwest, and duly met the vehicle with the equipment at a road junction on the Tanami Road. He was very surprised to find that the box of equipment had been bumping around in the back of the vehicle for hundreds of kilometres with no padding or protection. Moreover, he had had no training in using the logger, and was left to himself to figure out how to operate it.

He logged the four holes that were eventually drilled, but was very doubtful that the recordings meant anything, as the charts showed only a wiggly line with peaks and troughs only 2–3 millimetres in amplitude, and with no significant change from top to bottom of each hole. The equipment had recorded only background noise, and the logs were useless.

Fortunately Lyall's cuttings descriptions were excellent, and established the existence of a previously unrecognised but substantial basin of water-bearing sands and gravels entirely hidden by the sand plain at the surface.

While Lyall was occupied with the Portalogger, Ian and I parked the truck and four wheel trailer at Palmer Valley homestead, and detoured from Erldunda to visit Ayers Rock and Mount Olga, as this was Ian's first time in the Red Centre. We arrived at the base of Mount Olga on its western side just after lunch on a warm sunny afternoon, and Ian took a good look at the forthcoming climb. The climb, although it did not appear to be physically difficult, is an extreme example of the 'fly on the wall effect'. The rock surface is studded with pebbles and cobbles that provide good handholds and footholds, but is bare, smooth, steep, and slippery—one mistake and you go straight down. You can see all the way to the bottom, the top, and the sides, and the mental effect can be devastating.

Ian set off, followed slowly by me. I got about one third of the way up when my foot slipped, and some sort of subconscious message insisted that I give up and go down. Keen though I was to join the exalted few who have made it to the top, nothing could stop my legs, almost of their own volition, returning me to the ground. Later, when Ian met me back at the bottom, he said, 'That's one of the hardest climbs I've done.'

To our great sadness, Ian and Peter were killed the next year during their Colombian expedition, buried one night by an avalanche while they slept. Jennings Gorges in the eastern MacDonnell Ranges and Harverson Pass in the Reynolds Range are named after them. Harverson Pass marks the spot where Ian and Lyall first drove a motor vehicle across the Reynolds Range, although Gil Bowman from Pine Hill believed that many years earlier, a horse and cart had gone through. A few years after Ian was killed, his parents came from England to visit Australia, and I gave them a copy of our finished full-colour geological map of the Reynolds Range Region, which showed the pass named after their son.

7

RESCUES AND VISITORS

(1973–77)

In 1973 I stayed in Canberra to write up and produce maps, while Lyall and Alan Langworthy, from Russell Shaw's party, mapped around Alice Springs, and rented a house in the town to stay in between traverses. Lyall told me about a meringue he and his field hand Jeff Windsor tried to make. The CSIRO had recently invented a dried and powdered meringue mixture, and Lyall and Jeff, who liked to try new things on the campfire, bought a packet and took it out on traverse. One evening they opened the packet, followed the instructions, placed the mixture in a camp oven, and buried it in the ashes. After the specified time, they took the oven out of the ashes and lifted the lid. To their astonishment, the oven was empty; the meringue had completely vanished. They did eventually find it, stuck to the underside of the lid as a black coke-like mass of cellular charcoal.

/ / /

In 1974, Lyall, two field hands, Ian England and Linus Hellqvist, and myself mapped the Mount Peake and Denison map areas, north and west of the Reynolds Range. The area included two small tin mines, recently abandoned. We set up two base camps, the first near Anningie homestead on the Mount Peake sheet, and the second near Mount Denison homestead on the Denison sheet. Ian had been highly recommended by another BMR geologist who had worked with him previously, and he was efficient and resourceful. But we didn't hit it off. His outlook on life was so gloomy that I felt quite depressed at the end of our conversations, and so they largely ceased after two or three weeks. Our talk during traverse shrank to, 'Now we'll go over to that hill.' 'Turn left around the end of this dune.' 'OK, let's camp here.' and so on. At the end of the season, we parted on good terms, but did not work together again.

The Mount Peake map sheet includes Central Mount Stuart, a large hill of red sandstone about 300 metres high, and on the first traverse, we were able to drive both vehicles halfway up the gently dipping northern side of the hill before being forced to get out and walk. We were surprised to find an old truck tyre lying on the ground at the top, perhaps left behind by the National Mapping party that built the large cairn and trig beacon on top. This cairn has many notes in the crevices between the stones left by people who have climbed the hill. One I read was a rhapsodic account by a priest and young female follower who had climbed together before dawn, and wrote lyrically of the beauty of the view as the sun rose.

It rained two or three times that season, and one night when I was on traverse with Ian, our blankets got well and truly wet when the rain seeped under our swag covers. We dried them in the sun at lunchtime over some flowering acacia bushes. When we crawled into the swags that night, we were greeted by a strong smell of curry powder. The bushes we'd used were curry wattles, *Acacia spondophylla*. The curry smell lasted the rest of the week, a big improvement on the usual swag smell.

The canvas swag covers were supposed to be waterproof. When it did rain on traverse, which fortunately was not often, we pulled the swag covers over our heads and hoped for the best. The sound of rain falling on to canvas a few centimetres above your face is unforgettable. Heavy rain always got in. It flowed down the sides and then under the swag on top of the stretcher, and came in from below. We just put up with it, and dried blankets and sleeping bags the next day. Much later, the solution suddenly came to me—when it looked like there'd be rain that night, take the cover off the swag and rig it up as a low fly, with two corners tied to the vehicle and the other two to pegs in the ground. A central pole cut from a tree stopped any pooling of water on the fly. I never got wet on traverse again. As I lay in my dry swag listening to the rain falling on the swag cover overhead, I used to berate myself for having taken 25 years to think of this.

From 1994 on, we took one man tents into the field.

There are two very strange formations in the landscape near the abandoned tin mines, Anningie and Mount Allan, in the Mount Peake–Denison area. What appears to be a normal low ridge, a few metres high above the surrounding alluvial flat, turns out, when you reach the top, to be an annular ridge of pale crumbly rock, measuring some 10–15 metres in diameter and 1–1.5 metres deep. What makes it odder still is that inside the ring, at both places, is a pile about 5 metres across of round black boulders. The boulders are 1–2 metres in diameter, stacked two or three high, and composed of amphibolite. The whole thing looks like a pile of huge black eggs sitting in a nest. The boulders show a water level mark coincident with the top of the surrounding ridge, showing that the annular ridges hold water like a giant bowl after heavy rain. These nests of amphibolite eggs have puzzled me to this day. My theory on how they formed goes like this:

1. A column of hot mafic magma at about 1100°C intrudes into a vertical pipe in coarse-grained granite.
2. Heat from the cooling mafic rock changes the granite to a mass of tiny interlocking grains of quartz, feldspar, and mica in a cylindrical annulus next to the mafic rock column.
3. Metamorphism throughout the whole area at about 500°C converts the mafic rock to a very fine-grained interlocking aggregate of small grains, but has little effect on the previously affected granite of the annulus.
4. Erosion wears down the mafic rock and granite to their present level of exposure.
5. The granite in the annulus is harder, because its fine grain size means it weathers more slowly than the coarse-grained granite farther away, and so it forms an annular ridge around the mafic column.
6. The feldspar and mica in the granite of the whole area weather to clay, making the granite lose its coherence and become weak and crumbly.
7. Nevertheless, the abundant clay (65% of the rock) in the now-weathered annulus makes it markedly impermeable to water, and the annular ridge holds water for a while after heavy rain.
8. The mafic rock weathers most slowly of all, because its very fine grain size and widely spaced joints hinder entry of water into the rock, leaving it as relatively unweathered large boulders.

The Denison map area had wide sand plains covered in dense high spinifex with old, dry, and tall seed stalks that broke off and packed into every nook and cranny under the vehicles, including up and around the engine and exhaust manifold. We cleared out these stalks several times a day, but in the end the inevitable happened, and on the last traverse of the season, on a Friday 13th, Lyall's voice came through on the 6 p.m. skid with, 'We've burnt the vehicle.'

After I'd recovered from the surprise, Lyall gave me detailed instructions on how to find him and Linus the next day, by vehicle. This would not be easy, as the sand plain had numerous mallee gums restricting visibility. Lyall of course had the colour air photos of the area, but fortunately we had a set of smaller scale black and white air photos in the base camp, and I used those for the search soon after lunch the next day. I can still see, burnt into my memory, Lyall running through the spinifex towards us.

They had discovered the fire when Linus saw flames in the rear vision mirror. Lyall sprayed the fire extinguisher into the engine compartment, but after a few seconds the fire flared up again. Linus then had the presence of mind, and the courage, to remove the car battery from the flames so they could radio for help once the fire was out. He suffered painful burns to his hands. Eventually he extinguished the fire by smothering the engine with his leather jacket. Then they had to put out the grass fire that had sprung up in their tracks, using the blankets from Lyall's swag. The patch of burning spinifex extending for some 25 metres behind the vehicle caused several anxious moments until they managed to put it out.

With the damage to its electrical system, mainly the insulation on wires and cables, Ian couldn't get the vehicle going, so we towed it to our base camp, and fitted some new parts, such as the coil. The owner of Mount Denison station, Leo Martin, came down next day and rewired the spark plugs and distributor with household cable. It got to Alice Springs under its own power, then went down on the train to Adelaide.

Oddly enough, there turned out to be no way that BMR could compensate Linus for his burns or the loss of his jacket, which I thought was very mean and petty. Our supervisor Dick Dodson suggested, by way of an RFDS telegram that anyone could hear, some creative entries on his time sheet for that week. This provided some consolation to Linus for his injuries and losses.

In the following year, 1975, there was no systematic mapping, so we used the field season to visit the metamorphic rocks of the Musgrave and Tomkinson Ranges in the far north of South Australia with several geologists from the University of Adelaide and the South Australian Mines Department. We then took them on a tour of the MacDonnell Ranges, and after their departure, revisited problematical localities and mineral prospects in the Reynolds Range area. We made our way slowly and carefully eastwards along the valley of the Woodforde River, thereby crossing through the Reynolds Range. We met our host from 1972, Gil Bowman of Pine Hill, near the Aileron-Pine Hill road the next day, 6th of June. He was very surprised to hear what we had done, and told us we were the first to drive vehicles through the range along the Woodforde River valley.

On this trip, a beetle flew into my left ear while I was reading by the campfire late at night, and would not or could not come out. The wriggling and scraping of its clawed feet on my eardrum were excruciating. Fortunately Alan Langworthy knew what to do—he pulled out a bottle of peanut oil from the tucker box and poured it into my ear up to the brim. The wriggling and the pain slowly lessened as the beetle drowned, and finally ceased, leaving me with an earful of oil containing a dead beetle. I drained out much of the oil by tilting my head, but was deaf on that side for several days. Ten days later, the whole beetle came out, entombed in a cylinder of orange wax.

In 1976, the International Geological Congress took place in Sydney, and two of its field trips, one before and one after the congress, went through central Australia. Allan Wells led the earlier trip through the Amadeus Basin, and several others of us from BMR went along as guides and drivers. The trip included a two day scenic flight from Alice Springs to Ayers Rock, and then to Ringwood homestead to visit the gypsum body that BMR had drilled and Aimorn and I had logged in 1968. Apart from Allan, the guides were not included in the flight. Our job was to leave Alice Springs with pre-packed cold lunches for the entire group, and camp overnight at the Ringwood airstrip ready to meet the group.

We collected chicken and salad lunches from one of the Alice Springs hotels. They were all prepared ready to eat in hinged individual foam plates with lids, and packed in crushed ice in eskies. The dirt road to Ringwood was quite rough, so after breakfast at the airstrip we thought we'd better check that the lunches were all right. On opening the first esky, it was immediately clear that they were not. The contents of the esky—all the eskies—were sorted into a layer of empty foam plates floating on a slurry of chicken pieces, salad, crushed ice, and water. The beetroot in the salad had dyed everything—food, ice, and water—pink. We spent the morning skimming chicken and salad fragments off the ice and water and repacking individual pink lunches. They were all eaten in due course without complaint.

Between the two field trips there was a period of about ten days without visitors. They were all in Sydney at the congress, so I went back to the Reynolds Range with Jeff Windsor as field assistant to visit an area where Gladys and Andrew had disagreed in their geological mapping. Fire had burnt through the area only a few weeks before, and by the end of each day we were black with soot. There were still fires burning a few kilometres away on the spinifex plains northeast of us, and at night this was a real worry, because the fires were plainly visible. We knew they were a good distance off during the day, but at night there was no way to judge the distance to them. All we could see were orange flames in the blackness, and we were very mindful of possible changes in wind direction. We didn't get much sleep during that traverse.

The next two years, 1977–78, were spent producing maps and reports at headquarters in Canberra, and my only field work was to take part in some BMR geological field trips. One, in 1977, was called the Inspection of North Australian Geology. The trip was to show each other our field areas—Georgetown and Mount Isa areas in Queensland, and the Arunta and Pine Creek areas in the Northern Territory. We spent about a week in each area looking at key exposures.

Aimorn made me a beautifully simple and ingenious mosquito net to use in the Pine Creek area. It consisted of two items—a rectangular piece of muslin about 2.5 metres by 1.5 metres, and a thin cord about 2 metres long. The net and cord rolled up into a small package, and took only a minute or two to erect. To set it up, I tied one end of the cord to the midline of the piece of muslin about a quarter of its overall length from one short side, and tucked the edges of the muslin under the edges of the swag, either on the stretcher or on the ground. I then tied the other end of the cord to a tree branch or roof rack or scrub bar of a vehicle, and pulled the muslin up to form a tent over the swag. At bedtime, I untucked part of the muslin, crawled into the swag, and tucked the muslin back in place. I could hear the whine of frustrated mosquitos outside the muslin for a few moments before falling into blissfully undisturbed sleep.

When the North Australian trip was finished, I picked up a BMR vehicle in Darwin to ferry to Alice Springs. I set off about midday, but to my dismay after an hour or so, the alternator warning light came on and stayed on. I was driving with no charge going into the battery. There was nowhere to stop and have it seen to, and I had no spare alternator, so there was nothing for it but to keep going and hope I could reach the next town, Pine Creek.

After another hour or two, the engine began to lose power, and the vehicle went slower and slower. As I reached Pine Creek there wasn't even enough current to power the instruments, and the needles on the dashboard gauges slowly swung down to zero. I managed to roll into the town's only service station as the engine finally died. However there were no spares available in Pine Creek, and it looked like I was going to be stuck there for a while.

I went and sat by the road, hoping another BMR vehicle would come past, as I had heard in Darwin that a geophysical party was setting out for Alice that day. Sure enough, after about half an hour, Bruce Goleby of the Seismic Section came past, saw my predicament, and said he could tow me to Katherine, the next town south along the highway. So we set off, on one of the most hair-raising drives of my life.

I had assumed we would proceed along the road at a sedate and sensible speed, but Bruce simply got up to his normal 80–100 km/h. This was exhilarating at first, but things became frightening when Bruce's vehicle suddenly seemed to be reversing towards me. Naturally I braked quite hard, which gave Bruce's vehicle the appearance of going forward, so that I stopped braking, which led to violent tugs on the towrope and terrifying lurches to and fro by my vehicle. All that had happened was that Bruce had slowed for a few moments, but this had appeared to my eyes as if he had gone into reverse, while the rest of my brain knew we were going forwards.

This happened several times, until I learnt how to handle it. Eventually we reached Katherine, and mercifully, a new alternator.

8

GOLDFIELDS AND MULGA

(1979)

In 1979, Lyall, myself, and another BMR geologist, Bob Tingey, moved to Western Australia to work in the Youanmi–Sandstone area, about 500 kilometres northeast of Perth. It was a joint mapping project with the Geological Survey of Western Australia (GSWA), in the flattest but toughest country I ever worked in. Apart from numerous small gold mines, the area is used for raising sheep, and so, instead of a base camp with tents, we stayed in shearers' quarters. These are just about the most basic accommodation, but they have showers and a kitchen with cooking equipment and a wood stove, which makes things very cosy on a cold winter's night. The area is much farther south than the Red Centre where I was used to working, and cold rainy days and nights are common there. I also remember the area for an earthquake, an aurora, and a bad dose of flu that I caught at the Sandstone gymkhana.

GSWA's practice was to spread fieldwork over about six months, interspersed with two periods of about three weeks each when the geologists returned to Perth for a break. This did not suit BMR, for there was nothing for us to do in Perth, and two return flights to Canberra cost too much to justify. So our timetable was for me to work in the field for two months, return to Canberra for two months, then return to the field for the final two months. Bob and Lyall were to arrive in the field a month later than me, and do three months' field work each. This meant that when the GSWA geologists left after the first six weeks, Bob and I were still in the field. Because it was a joint project, GSWA

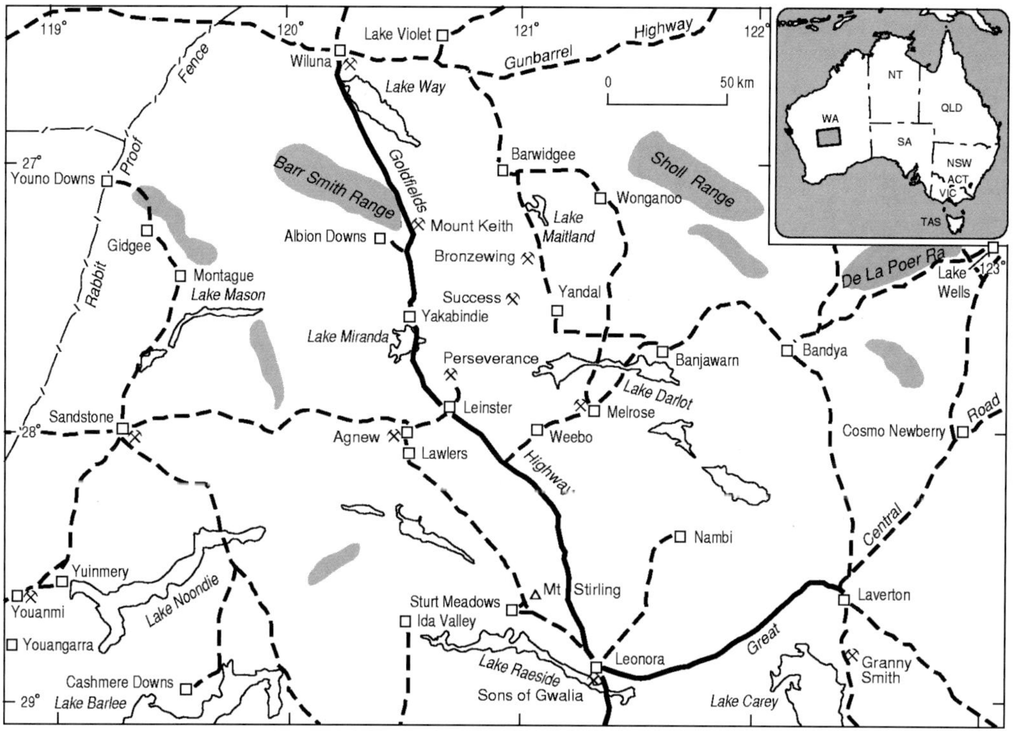

Locality map, northern Eastern Goldfields and environs, Western Australia.

had agreed that we could use their radio frequency and call sign at lunch times to talk to their geologists, who used the same frequency to report their daily positions to the GSWA radio operator in Perth. On the first day after all the GSWA geologists had returned to Perth, I called Bob at lunch time on the usual frequency. The radio operator in Perth cut in with a legalistic diatribe about our not being allowed to use the GSWA frequency or call sign, 'You are in breach of departmental regulations,' and so forth.

I had been doing exactly that for six weeks, with GSWA approval. After recovering from the surprise, I told Bob to switch over to our BMR frequency, and we used that for the next two weeks. In Perth on my way back to Canberra, the GSWA supervising geologist apologised and admitted that they had quite forgotten to tell their radio operator about us.

After about two months, my Landrover lost power, and took to spraying oil throughout the engine compartment. The oil was quite clean, which puzzled us greatly. After several days of this, Lyall and his field hand/mechanic, Dave Gregg, who were mapping on the adjoining Sandstone sheet, came down one evening to our base at Yuinmery, a few kilometres east of Youanmi, to have a look.

Abandoned township of Youanmi, WA, 1979.

With no workshop facilities, and only the trouble lamp to see by, Dave took the cylinder head off the engine, which revealed a large hole in one of the inlet valves. As well as causing the loss of power, this also explained the spraying oil. As the affected cylinder fired, the force of the explosion travelled through the hole in the valve, back through the carburettor and air cleaner to the outside, picking up oil from the air cleaner's oil bath on the way. Dave re-assembled the engine, and it started like a charm, although only on three cylinders.

The vehicle could not be driven back to Perth for repairs because a long drive with one cylinder out of action would unbalance and severely damage the engine. So BMR headquarters in Canberra arranged that a day or two later, around midnight, I was to meet a truck driver who was passing through Mount Magnet on the way to Perth. The truck driver knew of a loading platform at the abandoned railway station, but we discovered that it was about half a metre lower than the tray of the truck. Hunting around by torchlight, we found two thick timber planks to use as ramps, but they were of different lengths. We set up the longer one under the left front wheel of the Landrover, and the shorter one some distance nearer the truck in front of the right front wheel. I hopped in, put her in second gear low range, and gunned the engine. The front wheels drove easily into the truck, albeit on a slant, but the right back wheel only just made it, as the shorter plank dropped with a clatter into the darkness. We congratulated each other on our good luck, and I flew out of Mount Magnet the next morning for the mid-season break in Canberra. Transport to the Mount Magnet airport was by ute—in the back among the parcels.

The mulga forests south of Youanmi were so dense and high that my Landrover left a tunnel after its passing, with a carpet of smashed bushes and shrubs under the over-arching canopy. The damage to the vehicle was horrendous—the exhaust pipe was first broken, and later torn off together with the silencer—necessitating ear plugs for the occupants for the remainder of the field season—the rear brake line was ruptured, one petrol tank sprang a leak and had to be replaced, part of the scrub bars were torn adrift (but kindly welded back in place by the station owner, Philip Lefroy), one headlight was smashed, the throttle link came off, the rear door handle was torn off when we reversed into a tree, the side panels were squashed in, the front wheel arches were torn open as if by a tin opener, fanbelts lasted only a few days before breaking, and we lost the filling plug from the front differential. The mulga was unavoidable, however, as it stretched unbroken for many kilometres. The low outcrops we travelled to were invisible through the mulga until we were within a few metres of them, so we navigated by compass and distance. If we missed the target, we had the problem of guessing whether it was to the left or right of us. These mulga forests were the only places where I ever became lost in the field, for once out of sight of a vehicle, mulga scrub looks the same in all directions—this was long before the global positioning system (GPS) was invented. On two occasions I couldn't find the vehicle on returning from a walk, which produced momentary surges of panic, and only much yelling eventually re-established contact.

My field hand, Tom Baker, was cheerful and good humoured, did all the driving, fixed innumerable punctures, kept the vehicle going in spite of all the damage caused by the mulga, nursed me for a week in the shearers' quarters when I had the flu, and being by trade a chef, did all the cooking as well. At one stage we were stuck for a day or two in the Cashmere Downs shearers' quarters because of rain, so, while I did air photo interpretation, Tom cooked up some fancy dishes, including a large apple pie, just to pass the time. On our weekly trip into Sandstone for supplies, the GSWA and BMR parties had dinner together in the hotel. The publican was not keen on serving meals, and her husband even less so, being a landscape painter. Tom more than once ended up cooking the meal for all of us himself. After the field season, he bought and renovated an old hotel in Tasmania.

While I was in bed with the flu at Yuinmery, an earthquake occurred right under me. Its epicentre was at Cadoux, a small town 180 kilometres northeast of Perth, where it did quite a lot of damage, including breaking the town's concrete water reservoir. It began as a distant murmur in the southwest, which rapidly approached and became a roar, like a sustained dynamite explosion. When it reached the shearers' quarters, the huts rattled and shook violently for a few moments, and then the noise and disturbance died away to the northeast. The whole incident lasted only several seconds.

The aurora appeared one night in August as a motionless red glow in the southern sky decorated with moving yellow bands that constantly dimmed and brightened. Auroras form when charged particles from the sun—the solar wind—interact with the gases in the Earth's atmosphere; the red and yellow colours arise from electrically excited oxygen atoms.

Mulga forest (*Acacia aneura*) near Youanmi, WA, 1979.

9

PUNCTURES, MICE, AND WATER

(1981–83)

In 1981, BMR began remapping the Davenport Province, southeast of Tennant Creek. This is an area of very old sedimentary and volcanic rocks, deposited about 1,840 million years ago, and with virtually no petroleum potential. It did, however, have small abandoned gold and tungsten mines, and the largest tungsten mine, the Pioneer, was being re-evaluated while we were there.

To get into much of the Davenport Range, we often used a valley eroded on one of the mafic volcanic units. The number of punctures sustained from the fire-sharpened stakes in this valley was a major topic of conversation and source of irritation. One particular Friday evening when returning from traverse, my field hand, Ian Stepto, who had his partner Rose and their baby daughter Rhiannon living in the base camp, became more and more exasperated at our slow progress and the constant interruptions to fix punctures. Finally, as dusk was falling we came to a relatively flat area, and Ian decided to go for broke. He planted his foot and roared over the ground at 80 km/h. After two minutes we had four tyres flat on the ground, one very morose field hand, and an extra night's camping out. The entire following morning was spent repairing punctures, six in one front wheel, four in the other, and several in the back wheels. There were another 64 stakes in the six-puncture tyre and

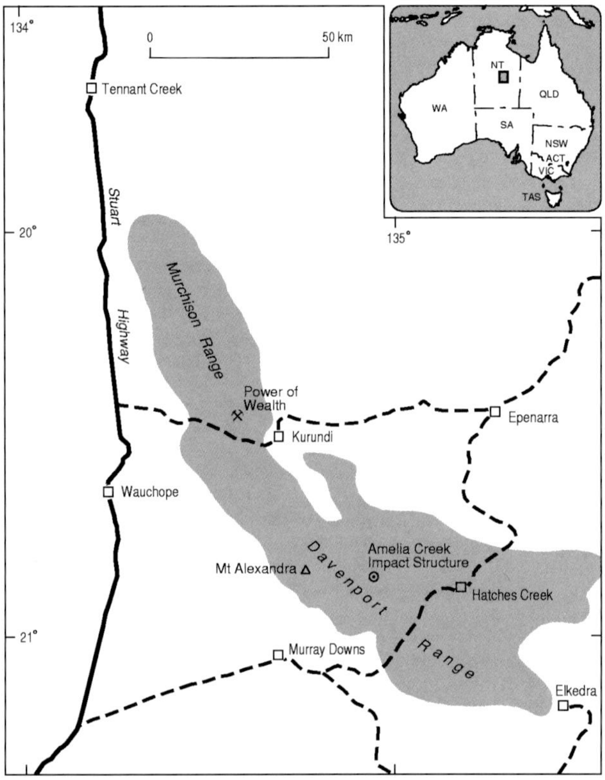

Locality map, Davenport Range and environs, Northern Territory.

48 in the four-puncture tyre that had snapped off before puncturing the tube, and all these had to be removed as well.

One day in 1982, the second year of Davenport fieldwork, my field hand Noel Young and I were returning to base camp. It was a Friday afternoon, the weekend loomed ahead, and we were bumping carefully but happily across an alluvial plain when the vehicle inexplicably stopped. A few years earlier, in 1977, BMR had organised a course over several Saturday mornings at the Canberra Institute of Technology on Emergency Vehicle Procedures, or 'What to do when the vehicle breaks down miles from anywhere', and I made quite detailed notes of what the instructors told us. After checking the obvious and most probable causes and finding nothing wrong, I told Noel about the CIT course and the notes I had made, and asked him, 'Shall we see if they're any use?'

Noel agreed, and we proceeded through the diagnostic checklist like student cooks following a recipe. In due course we came to 'Point 3: Rotor OK?' The procedure described was to check whether the distributor rotor button was faulty, thereby allowing the current from the high tension coil to leak away instead of going to the spark plugs. One person had to grab the plastic part of the rotor button with pliers, and hold the bare end of the high tension lead from the coil near the rotor button in his other hand. The other person then turned the engine over. Noel volunteered to hold the rotor button and high tension lead, and when I pressed the starter button, he backflipped diagonally into the grass.

After picking himself up, he inspected the rotor button with my hand lens, and duly found a groove filled with grease and dirt down its entire length. He cleaned out the groove with a knifepoint, and the rotor worked well enough to get us back to camp. I must admit to feeling just a bit smug at having taken notes and brought them along.

Graham Newman, the mechanic/field hand in our Davenport Party was a laid-back type, and the slowest, most relaxed, and most comfortable cross-country driver I ever encountered. We seemed to just tootle along between outcrops, but still got all the work done in the week's traverse. One morning we set off after breakfast, and came to a steep creek crossing. As usual, I got out of the vehicle to watch it make the crossing, and it went down the bank, across the stony creek bed, and started up the other bank. Half way up, the vehicle gave a tremendous lurch to one side, and an object like a small blue flying saucer flew off the roof and landed on a clump of spinifex. It turned out to be a bowl of milk pudding, one of two that we had made the night before and put on the roof to set. I had forgotten to eat mine, and the next morning it had travelled several hundred metres on the roof before being frisbeed off at the creek crossing. The bowl was sitting the right way up in the spinifex, the pudding safe and intact, so I ate it there and then.

Brown snake, Davenport Range, 1982.

Pebble mouse's pebble mound, Davenport Range 1982. Pop-up holes are just left of centre and in the upper centre, just left of the large cracked lump of rock. Scale (bottom left corner) 15 cm long.

In the Davenport Range, we almost discovered a new species of native mouse—a rodent, not a marsupial. During the mapping, we had all noticed carpets and mounds a metre or so across composed of fingertip-sized pebbles, and I asked Aimorn's colleagues at the Department of Zoology at the Australian National University what had made them. They referred us to the *Complete Book of Australian Mammals*, which described and showed the pebble mounds and the mouse that builds them, and included a map of their known distribution.

Nothing was recorded in the Davenport area, so I wrote to the author of the article, Darryl Kitchener at the Western Australian Museum, and found to our surprise that a mouse from one of the Davenport mounds had been caught only a few months earlier, and had proved to be a new species, *Pseudomys johnsoni* (Johnson's false mouse). My field hand Ken Mitchell was the first to find such a mound in the Mount Isa area of Queensland. I subsequently found many similar pebble mounds in the Mount Isa area, the Tomkinson Ranges of Western Australia, and the Mount Magarey area near Kulgera in the Northern Territory, all areas not previously known to be inhabited by this mouse or its relatives.

A pebble mouse with its pebbles (after the *Australian Museum Complete Book of Australian Mammals*, R. Strahan, editor, Angus and Robertson, 1983, p. 416. Printed by permission of HarperCollins Publishers).

No one quite knows why the mice build these mounds, or carpets, of pebbles. The mice live in a complex system of burrows beneath the pebble mounds, up to 25 mice living under one mound. They raise their young in nest chambers at the ends of short side passages. Condensation of moisture by the pebbles has been suggested as the advantage that their mounds offer.

An intriguing natural feature labelled on 1:250 000 topographic maps published by the Australian Government is the 'gnamma hole'. I had never seen one in central Australia, but they are common in the low granite pavements in the Eastern Goldfields of Western Australia. I had seen several in the Youanmi area in 1979, where they appeared simply as a lenticular or near circular hole in the granite pavement anywhere from 15 to 50 centimetres or more in diameter, often filled to the brim with water. My friend and colleague from the Youanmi mapping, Ian Williams at GSWA, had explained that the holes widen out at depth to a bottle shape a metre or more in width and depth. I didn't quite believe this at the time, but while on traverse in Skinner Pound in the Davenport Range, I came upon a gnamma hole exposed by a rockfall in the side of a cliff a few metres high in basalt. Sure enough, the portion remaining in the cliff was shaped exactly like a squat wine bottle—it was about 2 metres deep and half that wide, with a narrow open neck at the top. There were even a few centimetres of water in the bottom.

The bottle shape arises when the entire surface of a block of rock becomes a hard rind from the deposition of silica released during weathering of the core of the block to soft clay. Continued weathering at one particularly susceptible point can breach the casehardened skin, and the soft interior washes away, leaving an empty container or bottle, which eventually fills with water after heavy rain.

Over the years I long held a secret desire to get the names of my wife Aimorn and our daughter Alexandra on to a published geological map. I had unofficially named a peak in the Mount Isa area after Aimorn, and one in the Davenport Range after Alexandra. A cairn on each peak conceals a note to that effect sealed in a plastic sample bag. Family members' names for topographic features are not acceptable to government place name committees, but I eventually thought of a way around this. I was on good terms with the cartographic staff in BMR, and one was so good at hand lettering that he could make it appear indistinguishable from machine-printed lettering.

The first edition BMR geological maps showed not only geology, but roads, dams, etc., using a topographic plate supplied by AUSLIG. Very sneakily, I got my drafting friend to hand print 'Mount Alexandra' at the appropriate place, in exactly the font and size used elsewhere on the plate. The geological and topographic plates went off for fair drawing,[2] and duly came back with 'Mount Alexandra' now appearing as a proper topographic name on the Kurundi Region 1:100 000 Geological Map in the Northern Territory, and also on the ensuing Davenport Province 1:250 000 Special Geological Map.

2 Production of BMR's 1:250 000 maps was a two-stage process. Stage 1 involved tracing the geologists' line work from air photos (or air photo transparent overlays) on to large compilation sheets, reduction of these to 1:250 000 scale, and redrawing by hand. This produced an uncoloured map showing black line work and lettering, but it got the map out quickly. Stage 2 involved scribing of line work on to opaque films with a tool that rolled on three tiny balls and supported a cutting tip that scratched a smooth and clean line into the opaque coating. These were then transformed into positive transparent films. Lettering and symbols were typed into a machine that produced transparent plastic strips that the draftsman stuck in the required place on the appropriate film. A colour scheme for the rock units was then designed, sedimentary and metamorphic rocks by age, igneous rocks by composition. There could be up to 15 colour plates. At the printer, the plastic films were transformed into metal printing plates. If there were 15 colours to print, each sheet of paper went through the printing press 15 times. Back when maps were printed in this way, Australia's map printers were world leaders in this art.

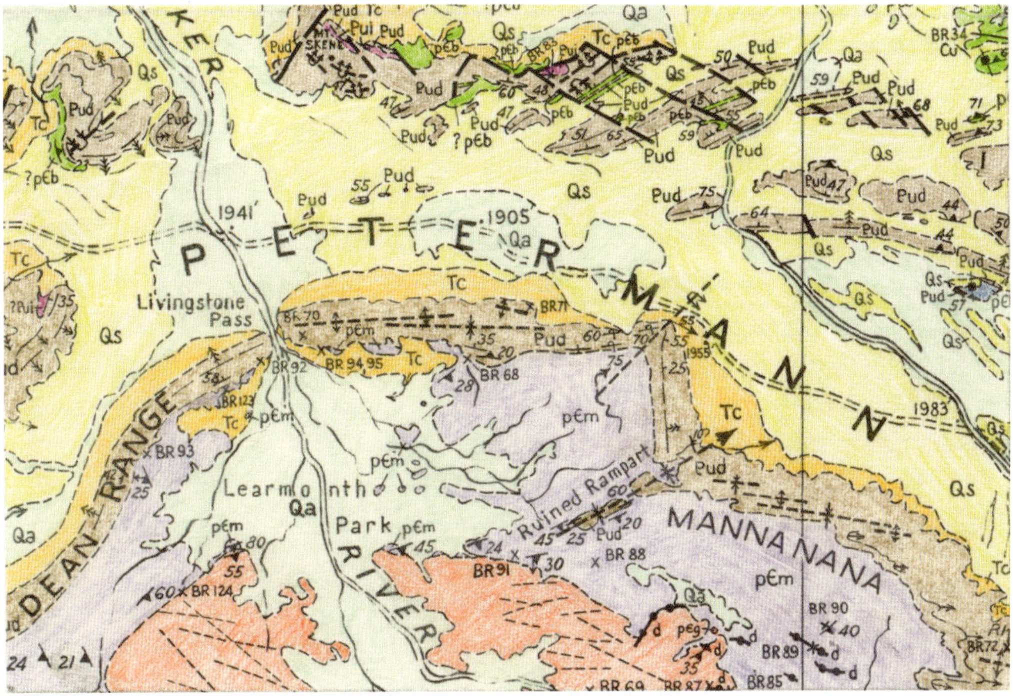

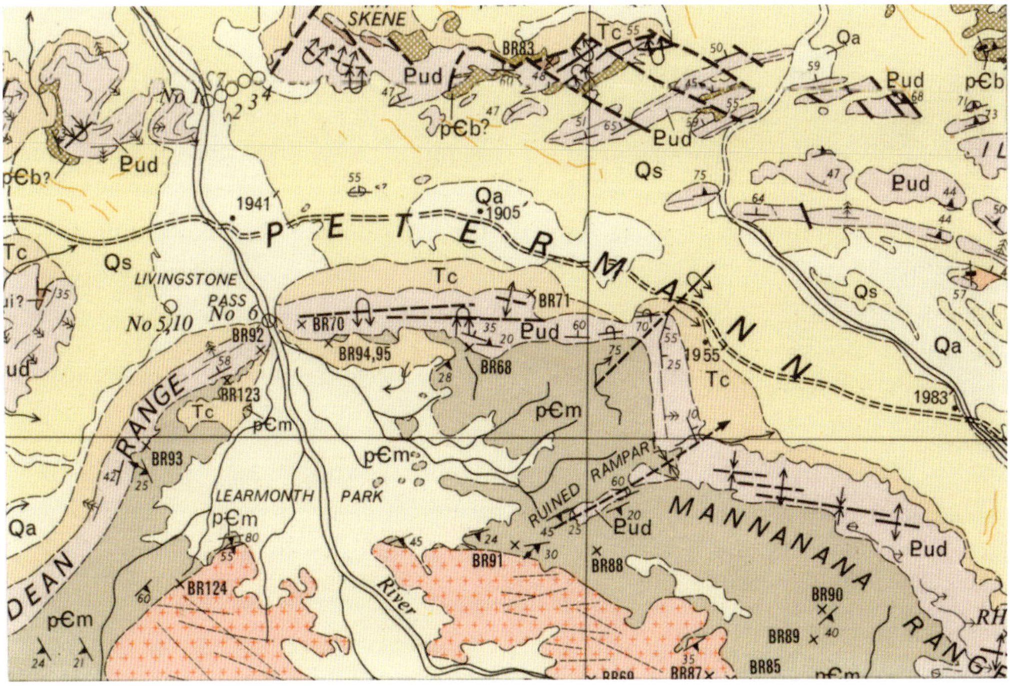

Example of a hand coloured Preliminary Edition of the southwestern part of the Bloods Range 1:250 000 sheet (1963, top) and the First Edition of the same area printed in colour by Mercury Press, Hobart (1967, bottom).

A few years later, I had an easier time labelling 'Mount Aimorn', as the relevant map was one of a set of 1:25 000 special maps of the Ballara—Mount Frosty area near Mount Isa that were hand drawn and then stored in digital form for printing on demand. Another drafting friend, also an excellent letterer, simply hand printed the name on the geological plate.

The Landrovers we were using for the Davenport work had two factory-fitted fuel tanks, one under each front seat. The water tank, instead of sitting up on top of a BMR-installed fuel tank in the cargo area behind the front seat, now sat on the floor of the cargo area. The outlet hose went through a hole in the floor and under the vehicle to a tap mounted on the rear chassis member, the lowest part of the vehicle. We used to worry that the tap could be turned on during creek crossings, and so we put a wire loop or ring through a lug to prevent it. As the wire had to be removed every time we wanted to turn on the tap, and then replaced after we'd turned the tap off, it was always forgotten and lost.

At dusk one evening late in the second field season, in 1982, we camped in the ranges a long way from any track, bore, or surface water. We unloaded the swags, collected firewood, went to the back of the vehicle to fill the billy, and there was the water tap in the 'on' position, with nothing coming out of it. We had small amounts of water in our drinking bottles, and in a canvas waterbag hanging on the front scrub bars, totalling about a litre. This would have to do us for cooking, washing, and drinking for that night and the next morning, and it being late September, the weather was hot—very hot.

The first thing we tried was to jack up the front of the vehicle, and a few spoonfuls of water came out, too little to be of much use. Next, we set out to try and disconnect the hose from the outlet pipe at the bottom of the tank, but instead we accidentally broke the whole pipe right out of the tank. This in fact was the best result, as it completely emptied the tank and we got about two litres, for the top of the outlet pipe stood up into the tank for a small distance. Fortunately we had placed the washing-up basin under the water tank and so caught the lot. There was just enough for two meals but no washing. The next day, when we got to a bore with good water, we drank until we could drink no more.

The whole saga occurred because I had left my pocket-knife on a creek bank near that same campsite earlier in the week, and so I had decided to go back and pick it up on the last evening. Getting to that campsite entailed several steep creek crossings, which was how the tap must have been turned on, leading to all the damage and trouble. But at least I got my knife back.

On one Sunday off, several members of our Davenport Party visited the abandoned Power of Wealth mine in the Davenport Range, and climbed down the shaft, some 15 or 20 metres deep, for a look around. The shaft ended at a horizontal tunnel that extended some tens of metres each side of the shaft and was filled with dust-covered miners' junk, such as shovels, rope, and a small dump wagon on rails that had carried the ore to the bottom of the shaft for raising to the surface. We also found a small dusty suitcase, which I picked up and shook. From the dull thuds inside I suspected it contained sticks of explosive, so I put it down unopened, and suggested to all that we quietly leave. I reported this to the Mining Inspector at Tennant Creek, and in due course he visited the mine and disposed of what did turn out to be old explosives.

Shaking the suitcase had been a very silly thing to do, as the nitro-glycerine in old dynamite or gelignite can migrate to the surface of the stick and form drops of liquid that may explode when disturbed.

One of our Davenport field hands was Bruno Zimmerman, who sometimes liked to cook up something out of the ordinary. More than once when I returned from a walk in the mid-morning, Bruno had set up the camp table and chairs and served up pancakes for a luxurious morning tea in the bush.

The final oddity in the Davenport work was to find that Bruno and myself had been photographed from the air during a colour aerial photography contract. We had colour photos of nearly all the Davenport Province before we started fieldwork, except for an area in the north. For my final two traverses, I used old black and white air photos, and some weeks later transferred the locality points where I had done the observations on to the new colour photos when they were delivered to the office in Canberra. I noticed one run of photos had been flown at the very date and time when Bruno and I were working in that area. So I looked at the air photo with a strong magnifying glass, and sure enough, found a tiny yellow rectangle, which was the roof of our Landrover, clearly visible on three overlapping photos. We could not actually see ourselves on the hillside, being too small for photos taken from 5,000 metres up. However I doubt that many geologists can say that they are in their own air photo in their own field area!

The next year, 1983, was an office year, when we wrote up our results and produced the maps of the Davenport area. BMR had purchased its first personal computers, and I learnt word processing while writing my Davenport report. The computer—there was only one in our corridor—was laughably slow; one could go away and have a leisurely cup of tea in the 20 minutes it took to save a page of text.

10

LARGE PARTIES AND UNIVERSITY STUDENTS (1984–89)

During the early 1980s, the Australian Government redirected BMR's function primarily to earth science problem-oriented research, instead of systematic mapping. In 1984 I moved to the Mount Isa area as part of one of these research groups looking at various geological problems there. Rod Ryburn was the party leader, and the field hands included John Pye, Ken Mitchell, and a newcomer, Tony Conroy as mechanic.

Mount Isa provided both the best base campsite and best traverse campsite that I ever enjoyed. The base camp was at Lake Corella, which in those days was full of water—now gone as the dam wall broke a few years later—and a haven for birds. Our camp was on short green grass right beside the water, and the local birds became quite used to us and wandered through the camp unconcerned. Two grebes built their floating weed nest only a few metres from the bank, and a bower bird used to raid our kitchen stores and spill oats, etc.

The new Chief of our Division had joined BMR a year earlier, and had ideas of running field parties more cheaply than previously. Instead of setting up a normal base camp with tents, a shower, a cook, etc., he instructed geologists and field hands to swag it in campgrounds in Mount Isa or Cloncurry, use the campground laundry and so on. They had started doing this the year before, but

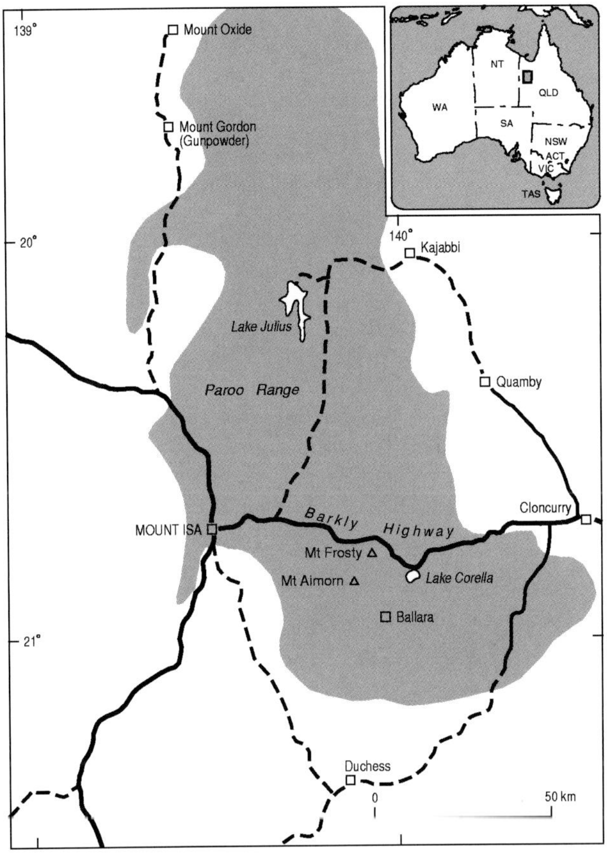

Locality map, Mount Isa area, Queensland.

found it totally unsuitable. Trying to lay out air photos, do correspondence, etc., on a swag in the open air is not practicable. Partway through the season, the party managed to smuggle in some camping gear with a drilling party that was passing through.

By 1984, our Chief had relented somewhat, and allowed a partial but discriminatory base camp to be set up, with tents for the geologists but not for the field hands, and a drafting caravan but no mess tent, so we had to eat out in the open. The field hands were issued small flies to sleep under, which kept the rain off when you were tucked up in your swag, but were of no use for anything else.

In 1985, we got a new Division Chief who was less stingy, and normal base camps resumed. In fact, they became very much larger than ever before; for three field seasons in Mount Isa we accommodated geology honours students and staff from Monash University for several weeks at the start of each field season. On occasions in these camps, 30 people or more sat down to dine.

The Monash students came out with us three times, until the Mount Isa project finished. They were noisy, and their taped music was worse. We needed a lot of organisation and logistic support, such as swags and food for twenty or more extra people, vehicles to ferry them to and from the camp to their mapping area, and frequent refilling of the camp water trailer as it was always running dry. On the other hand, it was fun to have so many youngsters in the camp, and we rather missed them when they were gone.

A year or two later, still in the Mount Isa area, I happened to work where two different travel allowance zones met. On some nights we would camp in one zone, and on other nights in the other zone. The finance clerk in Canberra had dutifully informed us that we would have to record our position, either as a grid reference or latitude and longitude, every night and show this position on the claim form for every night while camping on traverse so we would be paid the correct camping rate. So we did. The nature of the country and the camping conditions were the same, but we got more money in one zone, and less in the other.

On one traverse in the Mount Isa area, my field hand Ian Murdoch and I camped in a narrow rocky creek bed, to get away from the spinifex that covered the adjoining hills. In the morning when it came time to move on, the vehicle would not start, as the battery was completely flat. We jacked up one rear wheel and tried spinning it and then engaging gear to jolt the engine, but this did not work—the rear wheel simply stopped with an almighty jerk, and the engine did not turn over at all. Ian, an ingenious farm lad from south of Canberra, then removed the battery, emptied all its acid into an aluminium billy, and heated it to nearly boiling on the campfire, which cleaned the billy, making it look like new. He poured the fuming hot acid back into the battery, and the vehicle started effortlessly.

Mount Isa being one of the warmest areas I worked in, we sometimes met snakes towards the end of the field season as the weather began to warm up. On the very last day of one season, while I was walking back along a track to the vehicle after visiting the last outcrop, a small and indignant snake reared up and hissed most alarmingly at me before slithering rapidly away. More surprisingly, I met a snake during the cold early part of the field season. I had walked a few kilometres across country, and decided to take a sample of sandstone from an outcrop with many wide cracks in it. I hammered vigorously and noisily at the rock, broke off a suitable piece, to label and put in my rucksack. A large brown snake was coiled up in a crack in the rock a few centimetres from where I'd been hammering. The cold must have made it torpid, and it was my good luck it hadn't responded to the hammering.

Bush flies are a real curse in the outback, and we tried many brands of insect repellent to keep them out of our faces. I sometimes used to wonder what flies did at night or when it rained. I found the answer to the second question in the Mount Isa area, when I was caught out in a heavy shower far from the vehicle. I put on a poncho I carried in my rucksack, sat down on the side of a stony ridge, spread out the poncho like a tent, and waited out the storm.

After a few minutes, I realised that there were dozens of flies in a small twiggy shrub beside me—they were sitting motionless on the twigs and branches, waiting for the rain to stop. Colliding with a raindrop must be terrible for a fly. The flies had chosen only the most steeply angled twigs, and all without exception had their heads down facing the ground, and their bottoms up facing the rain.

Another time when I was caught in heavy rain far from the vehicle, I thought I'd be clever, and hurriedly took all my clothes off and stuffed them in my rucksack. 'At least they'll be dry when I put them on after the rain stops,' I thought. So I sat naked and cold in the pelting rain for 15 or 20 minutes. When the rain stopped and I had dried out, I opened the rucksack eagerly anticipating dry clothes, and discovered that they were sopping wet—the rucksack was of nylon mesh, and completely permeable. I had to put my wet clothes on, walk for an hour or more back to the vehicle, and then spend the evening drying them by the smoky campfire.

The next day it rained absolute torrents straight after breakfast, and we returned to base camp.

On another day when I took a walk by myself, I sat down by a small rock pool in Duck Creek to have lunch. It was a pretty spot in a small gorge, with gum trees, birds, blue sky and a warm sun—perfect for a midday break. One of the large orange and black hornets that often hang around waterholes came in for a drink while I was chewing. It flew over the pool for a while, hovering at different spots for a few seconds, and then made its final approach to the water's edge. But as it touched down, something went wrong, and the beautiful hornet crash landed. One moment this tiny lord of the air was showing off its aerial control and grace, the next it was an ungainly tangle of wings and orange legs and muddy sand. It scrambled back to its feet, however, had its drink and took off and away.

On another traverse in the Mount Isa area, while walking across a nearly treeless spinifex-covered plain, I saw in the distance what appeared to be an enormous pile of dead spinifex, some 4-5 metres across and 2 metres high. As I got closer, I could see a gap at the base of the pile. I looked in and saw that the heap was hollow. I stooped and went in, and found that I could stand up inside quite a large room, with a flat floor of grey sand mixed with dried kangaroo droppings and marked only with kangaroo footprints. The roos were using it for shelter from the heat of the day. Kangaroos do not build, so some kind person, perhaps a wildlife ranger, must have made it.

Much of the Mount Isa area is criss-crossed by faults, which are surfaces many kilometres in extent along which chunks of the Earth move, causing earthquakes. The rocks beside these faults are cracked and broken, and erode to form gravel or scree slopes of angular splintery fragments which make a delicious crunchy sound when you walk on them. In fact, I could always tell when I was getting near a fault because the sound of my footsteps changed to the crunchy sound.

In one such area, I was walking along and began to suspect that someone else was also walking nearby. So I stopped, but there was silence. I continued on, thought I could hear distant footsteps again, but again silence fell when I stopped. I went on again, but this time when I stopped, the other footsteps continued, and came nearer. Finally, a bowerbird appeared in a tree above me, and I realised that this was the source of the phantom footsteps. I scuffed one boot back and forth across the gravel, and the bowerbird immediately made the same sound. I did this two or three times more, and each time the bowerbird imitated it. I thought I'd try a different sound, took out my notebook and riffled the pages vigorously. The bowerbird imitated the sound with much enthusiasm, and we spent several more minutes at this game, it was great fun.

I happened to pick a beautiful traverse campsite for the final evening of one field season in the Mount Isa area. I chose the spot on the airphoto, and when we got there, it turned out to be far more scenic than I'd expected. I had intended to do a day walk, so I drove with Bob Skinner—an excellent field hand who had been with us in the Davenport mapping, with a pleasant nature and very reliable—to within sight of the campsite so that Bob would know exactly where we were to meet in the evening. The site was on a small peninsula projecting into the lake at Gunpowder, near the abandoned Mammoth copper mine. Bob then dropped me off at the start of my walk, and returned to find a way into the campsite. This turned out to be quite difficult, as the lake was surrounded by thick scrub and forest.

When I got to the site in the evening, Bob had parked the vehicle on luxurious tree-shaded grass right at the point of the peninsula, so that we were surrounded on two sides by clear deep water with ducks and other waterbirds swimming around a few metres from us, and on the third side by green grass and trees. The swim after the hot, ten kilometre walk was unforgettable.

In the north of the Mount Isa area, the mystery of what caused a disturbed ring of Cambrian limestone about 20 kilometres across near Lawn Hill homestead, some 250 kilometres north of Mount Isa, had been festering away for years in the back of the mind of one of our regular field hands, Ken Mitchell. Ken had been field hand for John Ferguson when he was studying the Strangways impact crater in the Northern Territory. Other geologists had examined the Lawn Hill structure, but had not been able to come up with any explanation as to its origin, although there was plenty of speculation.

During one quiet week in 1985, when all the geologists in the Mount Isa Party were out on traverse, Ken sneaked off to Lawn Hill, a day's drive from our base camp. I can still hear the barely suppressed satisfaction in his voice when he called us up on the evening radio skid and exclaimed, 'I've had a shattering experience.'

On arriving at the limestone ring, Ken had driven straight to its centre, and within a few minutes had recognised and collected dozens of rock fragments with a very distinctive conical striated fracture pattern called shatter cones. Shatter-coned rocks have never been recorded from volcanic eruptions but have been found at several meteorite craters. Ken thus concluded that Lawn Hill is also the site of an ancient extra-terrestrial impact.

Ken Mitchell, discoverer of Lawn Hill Astrobleme, Mount Isa area, 1985. Ken is holding a fragment of claystone showing striations caused by shrinking and swelling during seasonal wetting and drying of the rock.

The interesting thing about this tale is its lesson for geologists. In arid Australia, wide flat areas are normally covered with soil and alluvium. These conceal the bedrock underneath, so they tend to be avoided by geologists interested in the bedrock. At Lawn Hill the area inside the limestone ring is a nearly flat alluvial plain except for a low curved ridge of siltstone across its northern part. The actual centre of the ring is quite flat and apparently featureless and had not been looked at by geologists, who confined their study to the siltstone ridge and the limestone ring. Ken, being a layman, took the obvious course of going straight to the centre of the ring, and within a few minutes had the answer that trained but blinkered professionals had missed for years. Subsequently, Ken identified three more impact structures in the Northern Territory —Amelia Creek in the Davenport Range, Matt Wilson in Gregory National Park (which he identified jointly with Ian Sweet, a BMR geologist), and Cleanskin, which straddles the Northern Territory–Queensland Border.

In 1989 I was writing up and producing maps, and during that year I also attended the International Geological Congress in Washington. The congress was preceded by many field trips to destinations of tourist and geological interest, including a rafting trip along the Colorado River through the Grand Canyon of Arizona.

That trip lasted nine days, and was one of the most wonderful experiences of my life. The scenery was beautiful, magnificently so, the geological exposures perfect, the white water rafting through the tumultuous rapids exhilarating in the extreme, and the evening campsites on beaches beside the river were a perfect end to each day. My last game of volleyball was on one of those beaches. The most spectacular sight was where, a few thousand years ago, volcanoes had erupted in and near the western part of the canyon. The molten rock flowed over the canyon walls and down into the river, forming cascades of frozen black lava and lava dams that dammed the river back upstream for several kilometres before they eventually broke and allowed the water through again. One volcano erupted right in the river itself, but now only a cylindrical plug of solidified lava remains, standing straight up out of the water for several metres. It is called Vulcan's Anvil, and the many crevices in the side are packed with coins and other small items. These are offerings to ensure safe passage through the most dangerous rapids in the Grand Canyon, Lava Falls, 1.5 kilometres farther downstream. On the other hand, some believe it is dangerous—bad juju—to touch or place anything on Vulcan's Anvil.

11

FINAL DECADE

(1990–2000)

In 1990 I joined Andrew Glikson's party in the Tomkinson Ranges of Western Australia; these are the westward continuation of the Musgrave Ranges in the far north of South Australia. This project had started as a research project into huge masses, many cubic kilometres in size, of interlayered mafic and ultramafic rock with potential for platinum, nickel, and copper. BMR's function had changed back to regional mapping, so my job now was to map the metamorphic rocks that hosted the layered bodies. The other members of the party were Erwin Feeken, John Vickers, Julian Towns, and Garth Doblinger. Our base was the Aboriginal town of Irrunytju (Wingellina), a few minutes' drive from Surveyor-General's Corner, where Western Australia, South Australia, and the Northern Territory meet.

My first traverse was in the Champ de Mars area, visited and named by Ernest Giles in 1873. Early on the second day, a tame, friendly, and good-mannered dingo walked up and joined us. Julian, the dingo, and I did a half day walk in the afternoon up to the top of the Hinckley Range and down again, and he trotted along with us for the whole trip, at one stage posing gracefully on a boulder for photos. After we returned to the vehicle in the evening, he stayed around while we cooked and ate dinner. Though we wanted to, we fed him nothing, but he stayed anyway, and soon curled up on my unrolled swag and went to sleep. When it came time for me to go to bed, he woke and reluctantly gave up his spot, there not being room for two of us on the stretcher, and I found to my delight that my swag was beautifully warmed right through.

Double rainbow, Charnockite Flats, 20 km west of Wingellina, Tomkinson Ranges, Western Australia, 1990.

The next morning he followed the vehicle from stop to stop for a while, but eventually realised that no food was forthcoming, and left us. On returning to Wingellina several days later, Andrew was delighted when we told him about the dingo, he said, 'Oh I know who that is. It's Irrunytju. When I had the flu last year, he slept on my bed and kept me company for the whole week. He's been raised in Wingellina since he was a pup.' The next year we met Irrunytju again, now living back in the town, and very much the worse for wear—thin, weak, and scarred. A year later, he was dead.

Wingellina was strewn with rubbish, there was a lot of fighting and screaming at night, the houses had holes smashed in the walls, and the people had nothing to do. Petrol sniffing was common, and a sniffer attacked one of our vehicles with a sledgehammer. After the first traverse, I took to doing six day traverses instead of the traditional five days to reduce the length of time we had to spend in the town.

I had only two field seasons in the Tomkinson Ranges area. In the second season, 1991, we finished the Western Australian part of the work, but were unable to do any more work in South Australia, although we had been allowed

TOP: Irrunytju makes himself at home at traverse camp, Champ de Mars, Tomkinson Ranges, Western Australia, 1990. Julian Towns (mechanic) tries to claim his swag.

ABOVE: Irrunytju on swag.

Grass tree (*Xanthorrea thorntonii*) and author, with Mount Gosse in the background, Bates map sheet, Western Australia, 1991.

to work there in 1990. The local Aboriginals simply denied us access to their lands that year. This was a great pity, as the geology in the Musgrave and Mann Ranges is complex and well exposed. There is also something special about the landscape—the steep-sided pink rocky hills and ranges rising out of the silent red sand dunes under the blue sky have an air of mystery and timelessness, and I used to love climbing those peaks.

In 1992 I mapped the eastern end of the Musgrave region, with Erwin Feeken as field assistant. The two of us worked out of the small settlement of Kulgera, which is a truck stop with a motel and police station on the Stuart Highway about 18 kilometres north of the Northern Territory–South Australia border. I had done reconnaissance mapping of the same area in 1963 with Bill Woodyatt and his rabbit traps, and now 29 years later mapped it again, jointly with Nigel Duncan, a geologist with the Northern Territory Geological Survey. This time we had high quality, colour air photos, and mapped in considerably more detail. Erwin and I had similar tastes in music, and had no trouble deciding what to play at the end of our working day. Operas and symphonies rang out into the desert during the evenings.

Male bowerbird's display bower, with his 'chick-magnet' collections of white snail shells (two heaps) and green berries, near Mount Magarey, Northern Territory, 1992.

Back in Canberra, a radio salesman came to BMR with a portable VHF two way radio from New Zealand that he offered to us to try out. I took one for a day, and drove out for various distances from Canberra towards Braidwood. At each spot, I dropped my daughter Alexandra—then aged around 23 years old—at the roadside with the test radio, which had two aerials that had to be strung out horizontally for several tens of metres a metre or so above the ground, at right angles to the direction of the incoming transmission. Then I drove on for several kilometres, stopped, and transmitted to her from the vehicle radio. She got a few curious stares from passing cars while sitting by the road, and at one spot cows and horses came up from the adjacent paddock to inspect her. As she liked animals, this was a welcome distraction while waiting for the radio to come to life.

The portable radio was a disappointment; we could hear each other over only a few kilometres at the most, and towards the end of the day, the casing of the radio, and with it the battery contacts, began to fall apart. So we did not buy any. EPIRBs—Emergency Position Indicating Radio Beacons—became available soon afterwards, and the need for hand-held VHF radio transmitters disappeared.

In 1994, I moved to the Eastern Goldfields of Western Australia, shown on the map in Chapter 8, where we again had a joint mapping project with the Geological Survey of Western Australia. There were many gold and nickel mines in the area, and we did detailed mapping to provide basic data for the search for further deposits. The other geologists in our party were Dave Champion, Elizabeth Jagodzinski, and Patrick Lyons. At this time, the Canberra Institute of Technology had a training course for geological field hands, and we took three 19 year olds—Ben, Berlinda, and Veronica—for the ten week practical section of their course. Female field hands brought several new experiences, such as scent in the vehicle and being scolded by Berlinda: 'Didn't your mother ever tell you to take your feet off the table? Off!' she said with an imperious wave, during lunch one day on traverse. We were based at the Western Mining Corporation's No. 2 camp at the Perseverance nickel mine near Leinster, and the field hands were quite a hit with the miners.

Our vehicles in the Eastern Goldfields were equipped with cell-call radios instead of the old two-way radios, and these could be used as a radiotelephone. The procedure to make a call was extraordinarily cumbersome. One had to go through 11 steps of pressing various buttons, including a wait between steps 8 and 9 for a certain whistling-gurgling noise that indicated that the radio had made contact with a beacon hundreds of kilometres away. Quite often it did not make contact, and then we had to go back to step 1 and start again. At the end of the conversation, we had to tell the person we had called to dial 99 before they hung up so as to break the connection at their end, otherwise call charges continued mounting up indefinitely, or so we'd been told. After we had finished at our end, we had five more steps to do before the connection was broken at our end. By the late 1990s, satellite telephones superseded the cell-call radios, and were very easy, if expensive, to use, being just like an ordinary telephone in your vehicle.

Roads and tracks in the Eastern Goldfields often change as exploration turns up new targets, or mines become worked out and abandoned. One afternoon, Elizabeth, who was driving, and I were travelling at a good speed along a straight and nicely graded exploration track near the Mount Keith nickel mine, with the track showing clear and uninterrupted on the air photo. We were wondering why there was no sign of recent use, when, to our horror, saw in front of us a new drainage channel with sloping sides about a metre deep and several metres wide that had been cut straight across the track. All we could do was brake as hard as possible, but even so flew through the air for several metres before landing with a tremendous bang on the floor of the drain. All the gear in the vehicle was trashed, one of Elizabeth's wrists was injured, and my lower intestine felt as if it had been squashed in a vice. To our relief the vehicle itself was undamaged, and we eventually drove on, up the other sloping side, very shaken and chastened.

/ / /

Although I had never seen foxes in the Centre or Mount Isa area, we saw them close up in the Goldfields. On one traverse near the Bronzewing gold mine, I woke in the morning, dressed, crawled out of my one man tent and found a fox curled up in the grass beside one wheel of the vehicle, fast asleep. I called out to Berlinda to look out of her tent, so the fox woke of course, but instead of streaking away, it calmly and sedately strolled around the edge of our campsite and eventually into the scrub, with the occasional backward look. After breakfast, Berlinda found our plastic wash-up basin upon returning from a trip into the scrub—the fox must have taken it off during the night. Patrick had a small coffee pot, which he used to put on the campfire after dinner, and one evening a fox came up, snatched the pot from the fire, and ran off with it. Patrick immediately gave chase, until the fox dropped the pot.

I had a more pleasing animal encounter in 1997, when doing some reconnaissance work out of Leonora. I was then in the habit of going on an early morning run, and had set out at dawn to run along the Kalgoorlie road for a few kilometres. After leaving the town and heading south past the Sons of Gwalia mine and Mount Leonora, I was joined by a kangaroo, which hopped along beside me for several hundred metres. I think in the gradually increasing daylight it realised I was not a roo myself, and it eventually hopped away into the scrub.

/ / /

Songfa Liu was a geologist whom I met in 1994 when he was working at GSWA before joining the Australian Geological Survey Organisation. In 1998, Songfa and I began mapping the Leonora 1:100 000 map sheet, and were based at Clover Downs shearers' quarters just outside Leonora township. We did day traverses, showered and ate at the Sons of Gwalia miners' mess each night, popped in to the pub for an after dinner drink on the way back to our shearers' quarters, and generally led the life of Reilly compared to previous years of swagging it out in the open. By this time, compilation sheets had evolved into orthophotomaps—colour transparencies—roughly one metre on each side, of the assembled colour air photos of the area, with all air photo radial distortion removed, and, moreover, the Australian map grid built in. Songfa had loaded the orthophotomaps into a laptop computer, and as we moved the cursor over the image of the orthophotomap on the screen, the grid reference of the position of the cursor appeared at the bottom of the screen, correct to one metre. Each evening, we located our day's localities by comparing the air photos with the orthophotomaps in the laptop, and obtained the grid reference to the nearest metre, instead of the nearest 100 metres of our GPS instruments. Songfa's patience and kindness in guiding me in the use of our field computer never faltered.

A beautiful redback spider lived in her web under the wash basin at the Clover Downs shearers' quarters, but she never bothered us, nor we her. One day a large 15 cm centipede fell into the web. The redback was less than a tenth the size of the centipede, but she managed to kill it, fastening her fangs into it, and began sucking the juices out. Every morning when I went there for a wash, the centipede was smaller and the redback larger. She did not move for several days, till all that remained of the centipede was a thin dry shell a few centimetres long. She then cut it loose from the web, and we found the empty shell lying on the floor.

I returned to Leonora in March 1999 to finish the Leonora mapping, with Songfa scheduled to join me a week or so later. However, Cyclone Vance passed over the town a few days later, dumped a great deal of rain, and I was unable to go out on traverse. It was quite a surprise to see that while the cyclone was passing right overhead, and the wind was at its strongest, that it was not also overcast and raining all the time; on the contrary, overcast periods half an hour to an hour long alternated with periods of sunshine, and when we looked at the clouds in the distance we could see that they formed bars or strips separated by clear areas. The cyclone must have had discrete cloudy arms separated by clear areas spiralling out from its centre.

After a week of photo interpretation in the Sons of Gwalia office, I returned to Canberra while the country dried out. Songfa and I went back in August, and completed the mapping in October. It was the last systematic field mapping season AGSO conducted.

Leonora had its own police station, and one afternoon, while mapping right in the town, I saw from the ridge beside the town hospital that the police had set up a vehicle and breathalyser on the main road between the town and the Sons of Gwalia mine. They were breathalysing all the miners as they finished their shift and drove back into town. Did they think the miners had spent their shift drinking down in the pit? The following evening, Songfa and I had our regular drink in the pub at about 8 o'clock—Songfa a pony of beer and myself a scotch and ice—and then set out for our shearers' quarters. I had driven about 50 metres when the red and blue lights appeared right behind me; they must have been lying in wait beside the pub. Out came the breathalyser, and as it was about one minute since I had drunk my last mouthful of scotch, I blew a very high reading. This was most worrying—drunk in charge of a Commonwealth vehicle after working hours! However they said I could have ten minutes for the alcohol to leave my mouth, so I sloshed around as much saliva as I could. After four minutes—I timed them—they tested me again and I was below the limit.

Songfa and I often drove through the town on our way to the day's work. One morning, a pack of town dogs suddenly ran out of the front yard of a house across the street in front of me. Seeing the vehicle, all of them turned away except the one in front, who crossed well in front of me. But suddenly it did a U-turn and ran straight back under the left front wheel.

I jammed on the brakes, but the front wheel bumped up over the poor dog, and then the back wheel. I got out and found to my horror that the back wheel was now stopped right on top of the dog. I rushed back in, drove forward, got out, and rushed back just in time to see the dog die in front of me.

Leaning against the vehicle with my head in my arms, flooded with misery and shock, all the while I hoped no one from the town had seen what happened. Songfa, who'd had been following and seen the whole thing, went back to our base and got a carton to put the dog in. We bundled the warm body into the box, drove to the shire clerk's office on the main street, and told him what had happened. He was very understanding and sympathetic, and said he would

Sand goannas (Gould's monitor, *Varanus gouldii*) mating, Leonora area, Eastern Goldfields, Western Australia, 1999.

check on the ownership and dispose of the dog. The dog turned out to be a stray, but I was very upset. To this day I can see the eyes slowly closing in its blood-smeared brown head.

Later in the year I had a happier animal encounter. It was the last day of our mapping, and I'd been up in hilly country all afternoon, but managed to drive across the hills to the final rock outcrop for the season. I got out, picked up my hammer, took a few steps around to the other side of the vehicle, and came upon a pair of goannas mating. Goannas are monitor lizards that grow very large—up to 2.5 metres long.

I froze, and watched fascinated at the two animals bonded together. The male was using the 'one leg over' position on the back of the female, with his front legs clasping her middle. They were quite still, except that now and again the male would gently touch the top and sides of her head with his mouth, almost as if he were giving her a kiss. It was a very tender sight. Eventually, as I turned to go off on my walk, I found a third goanna on the ground right beside my feet—it seemed to be just as interested as me, and stood in silence watching the pair. They had finished up and gone on their way by the time I came back half an hour later. The spectator was gone as well.

My last field season was in 2000, back in the Red Centre for 11 weeks. I had finished up at AGSO the previous November, and was field handing for Dean Hoatson, an old workmate of many years, even though he was a Collingwood supporter. We spent several days northwest of Alice Springs traversing and sampling the Mount Hay massif, a high range of steep hills covered in large boulders and tall spinifex. We made camp one evening on the northern side of the range on the Burt Plain. This is a very pretty place, carpeted with Mitchell grass instead of spinifex, and one can see for miles with the range as a backdrop.

That evening, I thought about the coming end of my fieldwork days, and found that I was wholly resigned to it. My retirement the previous year occurred because AGSO had changed its objectives, and transferred its regional mapping function to the State and Northern Territory geological surveys. Some 90 of us, including myself, were made redundant. But Aimorn and I had no desire to move and work away from Canberra. All our friends and activities were there. The field season with Dean was a bonus, and eased the wrench of leaving AGSO. I'd had a wonderful life exploring the outback, hiking alone across hills and ridges—often where no previous exploration had been done—finding rocks and minerals that no one previously knew about, camping out, living and working far from cities and traffic and crowds, and I'd made life-long friendships. The fieldwork was soon to end, but there were plenty of other things still to do.

12

LAST THOUGHTS

I retired from AGSO on 30 November 1999, but returned the next day to start finishing several uncompleted projects. The previous month, GSWA had asked me to compile new editions of the Laverton and Leonora 1:250 000 geological maps, and write the accompanying reports. Moreover, after completion of the Leonora fieldwork, no one at AGSO had thought to program production of the Leonora 1:100 000 map. Finally, there was a report to write on the Lake Violet 1:100 000 map, which I had produced in 1997, but not had time to write up. So AGSO gave me a desk and a PC, and I went to work each day as if nothing had changed. I (with the AGSO cartographers) produced the Laverton 1:250 000 and Leonora 1:100 000 and 1:250 000 geological maps, downloaded the field data for all the map areas from AGSO's database of field observations, and wrote the two GSWA reports.

Towards the end of 2000, numerous changes to AGSO had started to appear. New people joined to work in new fields such as a tsunami warning system, earthquake monitoring, landslide, flood, and volcanic risks, coastal and climate change, geothermal energy, vegetation monitoring, 3-d modelling, and carbon capture and storage. AUSLIG transferred into AGSO in 2002, bringing its topographic mapping and geographic information functions, and AGSO changed its name to Geoscience Australia.

During 2001–03, I wrote up the Lake Violet work at home, and then in 2004 returned to GA on contract to work with a group preparing a new digital geological map of the whole of Australia. This involved generalising published 1:250 000 and 1:100 000 geological maps to 1:1 million scale, a job I was ideally suited for. When that finished in 2009, I joined GA's National Database Group and prepared outlines of Australia's major sedimentary and tectonic provinces, located and read the latest published research on each province, and entered it into a database.

One of the most satisfying things about geological mapping with BMR was its pioneering nature—we made the first systematic geological maps, wrote down the first descriptions of the rocks and resources, and worked out the first comprehensive accounts of the geological evolution of large areas of Australia.

Many things changed over the years. Base camps became more comfortable, especially the showers, which evolved from an unsheltered canvas bucket hanging from a tree to a corrugated iron hut with concrete floor (but no roof) and a hot water system. Kitchen caravans with freezers and fresh food became available. Numbers in the camps grew, and visitors, commonly supervisors and so on, became more numerous. Post-graduate students and faculty members attached themselves to us, and this culminated in the whole honours year class from Monash University coming out with us for three field seasons in the Mount Isa area. All these extra people meant extra logistics and paperwork.

Camping out on traverse became much more comfortable—we used gas lamps instead of making do with the vehicle trouble lamps, brought out lightweight camp tables and chairs, installed car fridges in the vehicles, and in the Western Australian goldfields in the 1990s we had one man tents, as it often rains there in winter.

The gas lamps used mantles of silk soaked in thorium and cerium salts, which had to be burned to ash to make them usable, and there were many tricks used to prevent the mantles from breaking as we bush-bashed across the spinifex each day.

Two-way radios gave way to cell-call radio telephones and then to satellite telephones, which meant that we could ring up our families where and whenever we liked. Our bosses liked them too, as they thought that they'd be able to ring us up from their desks in Canberra, quite forgetting that the phones had an on-off switch.

Field seasons became shorter, from five or even six months to as short as two or three months, which made a big difference to families with one spouse out in the field and the other back in Canberra. As the State Geological Surveys expanded, joint field parties became common, which sometimes led to ill feeling when field assistants doing the same work in the same party were paid different amounts.

Diagonal scrub bars at the front of vehicles were prohibited after about 1983 because they purportedly reduced visibility, and protective plating under the vehicles vanished because it became too expensive to fit; both these items previously saved the vehicles from a lot of damage. Use of the vehicles in scrub became unpopular with the clerks back at headquarters in Canberra because of the resulting damage to the bodywork; it became apparent that the geologists' priority—to do a proper job—had been supplanted by management's priority of not scratching the paintwork. As management never went out on traverse and saw what fieldwork required, this was understandable, but irritating.

Injuries were very rare; splinters, spinifex needles, and occasional burns were about the worst we had to endure. There were exceptions, however. In the MacDonnell Ranges near Alice Springs in 1964, I slipped when walking down a scree slope, and found myself sitting on the ground with a sharp pain in one arm. A stake of fire-hardened wood had broken off a bush, travelled under the skin for two centimetres, and then exited. Out of curiosity, I pulled the skin open and peered down the bloodless tunnel to check whether there was anything in there. To my surprise I saw daylight at the other end. After about half an hour I was able to stand up again, and found the stake lodged farther along in my sweater. Many years later, an x-ray found a small piece of charcoal in my arm.

The worst accident I heard about happened in 1963 in the neighbouring Henbury map sheet, when our Assistant Chief Geologist, John Casey, was visiting the field party. They had organised a campfire evening in the base camp, and at one stage Peter Cook and someone else got up to put a large log on the fire. The log tipped and threw Peter right into the fire. Fortunately he managed to do a sort of one-handed somersault, but his hand landed right in the coals and was badly burned. The thing that even Peter found amusing was that the music was playing and Johnny Cash was singing about falling into a fire at exactly the same moment.

Our lack of a proper procedure in the early days for finding anyone who went missing seems foolhardy and negligent now. This nonchalant attitude persisted until 1987, when a geologist in the Mount Isa area failed to return to the vehicle one evening. He was meant to camp with two other geologists, who then spent several hours searching for him, and at around 2 a.m. notified the Mount Isa police. The group had been working near the Mount Isa–Cloncurry road, and the missing geologist had simply walked to the road, thumbed a lift into Mount Isa, checked into a motel, and was tucked up in bed while the others searched for him in the bush. After this fiasco, the Division Chief set up proper search and rescue procedures, and made the preparation of rescue plans mandatory before parties departed for the field. By the early 1990s, we'd begun using GPS instruments, and the problem of getting lost disappeared.

I hardly ever encountered anyone out in the scrub. In the 1960s there was our encounter with the Aboriginal family at Bloods Range in 1962, and two years later, an incident near Alice Springs. I was about to open a gate on the boundary of the White Gums Dairy, a small property that supplied milk to Alice, when the owner drove up, brandished a rifle, and told me to clear off. 'I don't like prospectors,' he said several times. I tried to explain that I wasn't a prospector, but he wouldn't listen, and I reluctantly left without doing my planned observations on his property.

In 1979 on a rainy night in the WA goldfields near Cashmere Downs my field hand Tom Baker and I had decided to camp in the abandoned Youangarra homestead, and as we were drifting off to sleep two men drove up, also seeking shelter. They were pastoral inspectors from the Western Australian Department of Agriculture. In that same year, I found the name of a BMR geologist scratched into a rock on a ridge near Cashmere Downs, and later he told me he had worked there with a nickel exploration company before joining BMR. In 1985 at Mount Isa, I climbed a ridge and from the top could see a group of university students and lecturers on a field trip several hundred metres away, but did not go near them. In 1994 in the WA goldfields, our first traverse was with all four geologists together, and while looking at an outcrop a kilometre or so from one of the newly opened Mount McClure mines, a grim-faced man approached us through the scrub and walked past without stopping, grumbling something like, 'You can't even go for a walk without meeting someone.'

In 1999, near Mount Stirling in the goldfields, I came upon an unlocked car parked in the scrub, and a few hours later met the owner, a gold fossicker with a metal detector. He was working up on the hill slopes instead of down on the alluvial flats, for he said he did quite well up on the gravelly hillsides.

I recall with affection and interest almost everyone I worked and camped with, but a few stand out as really memorable. Allan Wells was one—I worked with him in 1963–64, and he was one of the most gentlemanly and lovable geologists I ever met. He treated his junior geologists as equals, and generally put people at their ease with his dry wit and relaxed style. He was a great confidence restorer at times of stress.

The other great BMR character was Tas Armstrong. He was a mineral separation laboratory technician who by plain hard work built up a stockpile of mineral separates large enough to keep the geologists who analysed the separates busy for several weeks. So Tas was free to go out into the field each year as camp manager, and set up the base camps of large parties. Tas's base camps were about as luxurious as a government-supplied tent camp could be, and were all set up for the geologists to fly in one day, prepare their traverse work, and go out the day after that. He came out with me one year as a technical officer/field assistant in the WA Goldfields, and our traverse camps were also out of the ordinary—breakfast was ready and laid on the table when I woke up, but woe betide anyone who put toast crumbs in the butter!

Tas trained up one senior field hand, Ken Mitchell, who then also set up large base camps. These two men took a lot of the logistic burden off the party leaders.

Living harmoniously with others week in, week out, on traverse and in base camp, was generally easy as long as we helped each other with the work, tried to think of the other person and not do things that annoyed them, and talked to the field hands about things that interested them, not about rocks. There were inevitable clashes of personality, and in those instances the party leader would try to arbitrate. If that didn't succeed, one of the disputants would usually defuse the situation by resigning. If the conflict was with the party leader, the other person just put up with it if they were on the permanent staff, or resigned if they were part of the temporary field staff. When the field party was too remote for resignation to be feasible, then one just put up with it.

Working with field hands required good will on both sides, as the geologist had to tell the field hand where to go many times a day. As long as we both did our part of the work, separately or together, all was well. When there was a puncture, we'd change the wheel together; when the vehicle broke down, we'd try and get it going together, or when it bogged we'd dig it out together.

When camping for the evening, while the field hand would begin the cooking, the geologist brought in enough wood for the whole evening's campfire, and set up stretchers for both. As the field hand did most of the cooking, the geologist washed up afterwards. In base camp, everyone had plenty to do—the geologists plotted up the week's work and prepared the next week's traverse work while the field hands prepared and restocked the vehicles for the next week. There were the occasional clashes of personalities, and when that happened the sharing of work and maintenance of respect for each other became even more important.

Some of the things our field hands did were odd, to say the least, and sometimes infuriating. One of our mechanics had been a racing car driver in

England, and there were times when he thought he still was. He and I went out on traverse after heavy rain had fallen a few days previously, and the air, ground, and vegetation were still very damp. On the first night when it came time to go to bed, I was astonished to see him carefully hang all his clothes on bushes beside his swag. When we woke in the morning all his clothes were wet through with dew. Having no others, he had to put them on anyway, with some grumbling that lapsed into silence. We spent most of that five day traverse in silence, as he was often moody and seemed uninterested in the work.

A traverse of silence with an unhappy field hand was just about the worst experience. I used to feel that the unhappiness was my fault, yet we still had to live and work together, and instructions to the field hand became the only conversation.

Another field hand had several particularly annoying habits, the worst being his preference for driving over the roughest and dustiest parts of the roads, instead of avoiding them. He also used to take off his only pair of socks late in the traverse week and put his smelly bare feet on the dashboard at lunchtime. I eventually had to ask him to stop doing that.

Another field hand/mechanic we had was a real surprise. On our first traverse together, we spent the morning on dirt tracks driving to the mapping area, had lunch by the track, and then it was time to go bush. To my puzzlement, my field hand was reluctant to turn off the track. I insisted that we do so, so he simply spun the steering wheel hard left, drove blindly off the track and straight over a hollow tree stump of fire-sharpened spikes, immediately and thoroughly puncturing the rear tyre. I could scarcely believe my eyes. He then explained that he did not like off-road driving. I could now scarcely believe my ears. I was forced to rearrange the traverse so that all walks—which I did alone—began and ended at a track.

He redeemed himself partly when one evening I trod on and flattened my shiny red Swiss aluminium water bottle. I thought it was ruined, but he poured in a few millimetres of water, screwed on the stopper, and placed it at the edge of the campfire. In due course the bottle expanded to something like its original size and shape. He removed it from the fire, slowly undid the stopper, and gave it back to me. The paint was gone from one side, but the bottle was now perfectly serviceable, and lasted for all my subsequent field seasons. Even so, we did not go on traverse together again.

Cooks seemed to cause more trouble than other field hands. On one occasion, the party leader gave me the job of finding and hiring a replacement cook through the Commonwealth Employment Service in Alice Springs.

I interviewed about half a dozen applicants one Saturday morning, and chose the most experienced, who had cooked on cattle drives. He turned out to be very pro-apartheid, and lectured us at meal times with racist theories and why apartheid was necessary. We found this very discomforting, but did not know enough about genetics to counter his assertions. He also boasted about the rough and tough life in droving camps compared to our soft geological camp life—how he used to get up every day at 4 a.m. and 'tailed the horses', and how 'everyone ate a pound of thick steak for breakfast'.

He refused to sleep in any of our base camp tents, and unrolled his own swag each night down on the bank of the dry riverbed beside our camp. Despite his bragging, he found lighting the bottled gas-powered kitchen stove troublesome, and after one particularly violent explosion, left in a huff after only a few weeks on the job.

Another cook I hired had been a chef in the Defence Department in Canberra. She was pleasant and caring, and cooked delicious meals for our very large party. But after a week or so, it became clear that all was not well. She began to treat the field hands badly, as 'other ranks', whereas she curried favour with the geologists—the 'officers'—by giving them unexpected presents of cake and other treats as they were departing on traverse. She refused to clean up her pots and pans after preparing a meal, because chefs didn't do that. The field hands soon refused to help with the washing-up in any way, and one or two geologists had to do the lot. It was a great relief when she resigned after several weeks' of fieldwork. The field hands and camp manager cheered loudly when they learned that she had gone.

An extraordinary coincidence one field season proved that crime does not pay. Several weeks after the start of the season, it became clear that some of our tools were missing, but no one would admit to knowing anything about them. By the time we suspected enough to contemplate action, the helicopter had arrived for its ten-day contract.

Keeping the helicopter flying was a full time job, with evenings occupied by the planning of the next day's flying. One morning, the helicopter mechanic drove off to the nearest town to buy some stuff for the helicopter, and the pilot took off with the party leader and myself for the day's traverse. Within a few minutes the rev counter began to play up. By good luck, we were flying near the road to the town, and the pilot managed to catch up with the mechanic and surprise him by landing right beside the road in front of him.

While they examined the helicopter, we hung around the vehicle, and eventually noticed a large parcel in the back addressed to our party mechanic's wife. We found we could barely lift it, but when we did so and shook it, it was obviously packed with metallic articles. With our suspicions rising rapidly, we returned to camp, the day's flying aborted. We reported the matter to the police, and they opened the parcel in the presence of our mechanic. It proved to be full of the missing tools, plus a lot of others our mechanic had helped himself to the Department of Supply store in Adelaide. He'd taken them all the way to our field area, and only been discovered because of a faulty helicopter gauge. He got off with a warning, and later became a successful mechanic and hot-rod tuner.

Another time, unknown to me, the cook had previously worked as a mechanic, and the mechanic as a cook. They travelled together in the truck on the way to the field, and had fallen-out by the time we arrived at the field area. This coalesced into much argument between them over whether to use super or standard petrol in the camp generator. After a few weeks, the ridges between the grooves that held the rings in the piston in the generator engine partly melted, and the engine stopped. We thankfully abandoned the fancy fluoro lights in the tents and went back to using traditional gas lamps with mantles, which gave a much better light and a much quieter camp. The piston became an ashtray for visitors in my office.

A geological survey cannot operate without draftsmen and women—now called cartographers—who prepare the final geological maps that the public sees and uses. The cartographic staff of BMR-AGSO became world leaders in the production of high-quality geological maps, and won many awards at cartographic gatherings. Of the many I worked with, two particularly stand out. Andy Mikolajczak, a field draftsman with us in 1963, could do the best hand lettering I ever saw, and could make it indistinguishable from machine-printed lettering. Diana Pillinger who had trained as a photo-interpreter with the Royal Air Force, knew what geologists were doing with their photo-interpretation, and understood what they had to try and show on their maps. She was a fast worker and meticulous letterer, and could lay out the legend and diagrams surrounding the main map in a very logical and yet aesthetically pleasing way. She won a Departmental Secretary's Achievement Award in 1992 for her series of 22 special maps of the Mount Isa region.

/ / /

Separation from one's family was the worst aspect of the long field seasons. We wrote letters that could usually be posted or received once a week with the supply run, and we rang up from public telephones whenever we had the chance. The RFDS telegram session could be used for occasional personal telegrams if necessary; the sender paid for these, but we could not say anything very personal, as everyone listening in to the session heard everything we said.

The long field seasons—four to five months—of the 1960s did cause a great deal of loneliness for married geologists. Those without children quite often took their spouses into the field, and in 1964, Allan Wells' wife Sally, with a toddler and newborn baby were in the camp at Ross River. Once the children started school, however, spouses could no longer go out to the field.

Around 1968, things came to a head. When the nickel boom led many BMR geologists to join mineral exploration companies, staff numbers and morale at BMR began to decline. Field geologists still working at BMR realised that they now had some leverage—the threat of more resignations—to force management to listen to their request for shorter field seasons, or at least mid-season breaks, and eventually this came to be so. Even then, the break of ten to fourteen days in Canberra had to be justified in such terms as 'consultation with supervisor', 'discussion and map supervision with drafting staff', and so on.

An unexpected discovery I made was that over the years I had, without knowing it, become quite possessive of the areas I had mapped. I realised this in 1990, during a break from fieldwork in the Tomkinson Ranges. We drove from Wingellina to Alice Springs through the Musgrave Ranges via Amata and Mulga Park, but decided to return through Ayers Rock and the Petermann Ranges. As we neared Docker River, I diverted from the main road and drove a short distance into Learmonth Park and took a look at my old 1962 stamping ground. At that time my field hands and myself had been the only people living and working there, all the Aboriginals having moved away to stations, settlements, or missions. In the intervening years, Docker River settlement had been established and a road graded through Learmonth Park to smaller settlements near the Pottoyu Hills. Now, as I looked at the steep quartzite ridges enclosing the Park, and the magnificent Ruined Rampart to the east, I felt a strange but strong feeling of ownership.

'This is all my land' I thought to myself: I had lived, worked, and climbed all over it for weeks, made the first systematic geological observations of it, racked my brains over its geological structure and evolution, written a report about it, and in my mind it had become mine. I have to admit that I rather regarded the Aboriginals who had moved back there in the late 1960s as usurpers, even though I knew that they were the first people there. So I have to share 'my land' now.

Some years later again, in 2005, Allan Wells and I drove along the Docker River–Ayers Rock road while returning from a trip to Lake Christopher and the Rawlinson Range, where we had tried unsuccessfully to locate Lasseters Reef. I decided to have another look at Learmonth Park, and managed to find the traditional owner, Nookety, at Docker River. I explained that I had lived and worked there some 40 years earlier, and he immediately gave us permission to go in to the park.

We drove several kilometres in and had a good long look. For Allan it was his first time, for although he had mapped on the Rawlinson sheet in 1960, he had never been into the park, which is on the adjoining Bloods Range sheet. It had transformed and was thickly vegetated, in contrast to the drought-stricken landscape of 1962.

I found the clean, clear colours and the sharp harshness of the central Australian landscape very beautiful, and its silence and stillness made it seem almost timeless. In my first few years, when reaching the top of a hill and seeing the dunes and mulga stretching far away to some distant peak, I often used to think of Keats' lines:

Then felt I like some watcher of the skies
When a new planet swims into his ken;
Or like stout Cortez—when with eagle eyes
He stared at the Pacific, and all his men
Looked at each other with a wild surmise –
Silent, upon a peak in Darien.

It is easy to rhapsodise on top of a hill when one can just walk back down to a convenient supply of food, water, and transport. In contrast, the ability of Australia's first inhabitants to survive and live off the land in such a landscape is truly remarkable, for it is totally unforgiving of any errors.

There are clear, moonless nights out in central Australia when you can see half the universe ablaze. On other nights you can read by the full Moon, sometimes with a magpie warbling away in the distance, or best of all, the screams and shrieks of curlews. Over time, you see the movement of the night sky and the motions of the moon and planets, meteorites and satellites against the stars. Many changes take place over the years as droughts give way to good seasons; bare red sand with a few blackened sticks changes to parkland of shrubs and grass teeming with animals, red rocky slopes change to green as the mulga comes back into leaf. These reveal how active the desert of the Red Centre really is.

I loved working in Australia's outback. With no television, newspapers, or internet, you are free from the worries of the world, and when you are prepared for any eventuality and confident that you can cope with emergencies, life is simple and fulfilling. Navigating across trackless terrain is exciting and rewarding in itself, especially when you find an inconspicuous target in thick scrub. Walking alone for a whole day in that still, silent land, observing the rocks and recording your observations, is very rewarding. When you find an exposure of real complexity, where our planet has laid out the results of her complex processes of heating, cooling, and deformation, processes critical to our understanding of the Earth's geological evolution and the resources we need, life is very satisfying indeed.

APPENDIX I

HOW AND WHY FIELD GEOLOGISTS DO THEIR WORK

Our planet Earth supplies all our needs, whether dug, pumped, grown, pastured, or breathed. High on this list are oil, gas, coal, minerals, and water, and most nations have set up government geological surveys, beginning in 1835 with the Geological Survey of the British Isles, to find these materials. The search begins by making earth science maps and reports, for they provide the essential raw data. For Australia's first 150 years, the individual colonies—and after Federation, the states—gathered and published these data. Australia's national geoscience body began in 1946 as the Commonwealth Bureau of Mineral Resources, Geology and Geophysics (BMR), a government organisation with its headquarters in Canberra. It was part of the then Department of Supply and Shipping, and its chief functions were to conduct geological and geophysical surveys of Australia. In 1992 it was renamed the Australian Geological Survey Organisation, and in 2001 Geoscience Australia.

Geological field parties from the BMR worked mainly in the northern half of Australia and fieldwork took place in the winter. The summers were too hot although I did have to do one summer's fieldwork, described in Chapter 5, because of financial pressures. Each year, meetings late in the calendar year set up the next year's program: which areas were to be mapped, how many parties would do it, the personnel in each party, the number of vehicles, and so on. The field parties set up base camps in their assigned areas, from which geologists did weekly traverses in Landrovers. The base camps consisted of tents for messing and sleeping, plus washing and lavatory facilities, which ranged from none, to hessian or bough shelters, to corrugated iron shelters with concrete floors, depending on what the party leader wanted. Where the base camp was near a vehicle track, it would usually have a drafting caravan as well, where a field draftsman would trace the geologists' line work from air photos on to large compilation sheets. Back in Canberra, these were reduced to final scale, and the map redrawn from them.

Over the centuries, the study of the Earth has revealed that many deposits of metals occur in certain rocks or associations of rocks, but not in others. If we are looking for platinum (for our car exhaust systems for example), we don't look in limestone. Lead, on the other hand, is quite often found in limestone. Other metals, such as copper, occur in lavas that were poured out on to ocean floors, and this has been seen happening in present-day oceans, where openings on the floors of the Atlantic and Pacific oceans eject clouds of particles of copper sulfide (the famous 'black smokers'). If we can show that solidified lavas, or volcanic rocks, erupted on to a former ocean floor, exploration companies will feel optimistic about looking for copper in them.

Similarly, petroleum occurs in commercial quantities only in sedimentary rocks, never in igneous rocks such as granite or basalt, nor in metamorphic rocks such as schist and marble. A geological map immediately indicates in broad terms where and where not to look for petroleum.

Underground water is stored in the tiny spaces between the mineral grains making up rocks, or in networks of cracks in rocks. Large supplies, such as in Australia's Great Artesian Basin, occur in deeply buried sedimentary rocks, but in many places adequate supplies occur in aprons of alluvial gravel and sand around hills and ranges. A geological map must show the distribution of these superficial sediments, as well as the more complex hard rocks below.

Modern society also needs industrial minerals, such as clay for making bricks and crockery, limestone for cement, sand for glass making, abrasives for grinding and cleaning, as well as construction stones such as granite and basalt for buildings, paving stones, and bench tops. The piers at the ends of the Sydney Harbour Bridge are made of granite. Melbourne Grammar School, Pentridge Gaol, and St Patrick's Cathedral in Melbourne are all built of basalt.

Geology is the study of the Earth and how the Earth works. Observations many centuries ago of fossilised sea shells high up in mountains led people to ask how they got there, and present day geological knowledge stems from that. Rocks undergo many and varied physical and chemical changes, from the

Geological field boots—good quality was essential.

atomic to the continental scale, and at depths ranging from the Earth's surface to hundreds of kilometres down. Rocks change their shape from being squeezed, folded, and broken, and their constituent minerals react with each other and form new minerals. The results of these processes become frozen into the rocks as they cool, and their study reveals the processes that go on in and on the Earth. Some rocks can be quite beautiful, which makes them very attractive for use in building. In short, rocks have complicated histories which are fascinating to decipher.

Geological mapping involves walking across the countryside and looking at rocks. In many parts of the world, rocks, such as granite, schist and sandstone, are weathered to clay, sand, gravel, etc., because of chemical attack by air and water. The unweathered rocks become partly to wholly concealed by their own weathering products, and these are periodically washed away to form alluvial plains and sandy deserts. Geologists map out both fresh rocks and superficial cover, but the fresh ones are more complex and so require more investigation. They preserve a much more extensive and complicated story than do the cover sediments, and create the need to find and study exposures that reveal these complications, such as the relationships of several rock types.

Until the 1930s, such critical exposures or outcrops were found by direct observation, i.e., by looking at mountain or hill or valley sides and cliffs, where there was some vertical relief. In much of Australia the land is flat and relief generally low, so good exposures are hard to find. In 1929, Cecil Madigan at the University of Adelaide took the first air photographs in Australia, using a large camera mounted over a hole cut in the bottom of an aeroplane's fuselage. The air photos quickly solved several puzzles in the area west of Alice Springs, where the first geological workers had thought that a small number of continuous sedimentary formations extended east–west for hundreds of kilometres. The airphotos showed that there were more formations than first thought, and that they were not continuous, but tapered off in places, while in others they built up from nothing to a thick mass.

Subsequently, the Royal Australian Air Force and private contractors took black and white air photographs of the entire continent, initially at about 1:46 500 scale, and later at about 1:80 000 scale. Australia's first topographic maps of much of the northern half of the country stemmed from these photographs. When BMR began mapping northern Australia in the late 1940s, they had a ready supply of air photos. Each field party took with them the air photos of all the areas they were to map, and before every traverse, each geologist would examine the photos for the coming week's mapping. We used a stereoscope for the examination. A stereoscope is an optical instrument that combines the images from two overlapping photographs into one, thereby revealing the scene in three dimensions. A pocket stereoscope consists simply of two low-power magnifying lenses mounted side by side in a frame at the average distance between the centres of a person's eyes. The frame has legs that when unfolded hold it a fixed distance from the photos so as to keep them in focus. The viewer lays the photos down so that they overlap, with part of the photo under one lens, and the corresponding part of the other photo under the other lens. A table stereoscope has two inclined mirrors set apart so that successive photos can be laid side-by-side instead of overlapping, which makes them easier to see and draw on. Two prisms at eye distance apart collect the light from the mirrors and reflect it up into the viewer's eyes. Viewing everything on the photos in three dimensions, we could see any critical exposures instantly. These included places where rocks of different trend had been brought or faulted together, or where a younger sedimentary sequence overlay the eroded edges of an older sequence at an angle (forming an unconformity), or where rocks that began as hot molten magma had been injected into adjoining solid rocks. So we knew straight away where to go to examine useful exposures like these, and find out what the rocks and their relationships were. The air photos also showed where the best exposures of single rock types were, so that we could find out what a rock unit was made of before going to the complex exposures, where different rock units were present. The air photos also showed areas of overly rough or impassable country such as impenetrable scrub, cliffs, etc., that we would have to avoid. After deciding which exposures to visit and circling each one on the air photo with a grease pencil (a soft, coloured pencil that writes on virtually any solid surface) we clipped together all the photos for a planned week's traverse so as to make a photomosaic of the area, and then connected up the exposures to be visited and also the best route—the 'good going'—between each one. We would do this and lay out in advance a whole week's work, or traverse, before setting out.

Once on traverse with a week's supply of food, water, and fuel, the field hand drove and the geologist told him or her exactly where to go, sometimes every few minutes or so if in hills or irregular sand dunes. The most important task when driving between exposures or outcrops was to avoid punctures as much as possible—both front wheels had to be aimed so as to avoid any sort of bush, shrub, or dead splintery wood. Driving across country occupies one's complete attention.

Punctures were a curse when driving across country, and for a time in the 1970s they caused more than the usual annoyance, when vulcanising patches failed to stick to the inner tubes, apparently because of a manufacturers' change in the nature of the rubber of the tubes. This was eventually solved when someone found by trial and error that a repair would hold if you stuck the patch on the tube with rubber cement and then vulcanised it as well. In due course, stick-on patches made vulcanising patches obsolete. The only thing worse than punctures was vehicle breakdowns resulting from broken springs, which were the most common failing of the Landrovers of the 1960s. These Landrovers, designed for farm use in Britain, were never intended to cope with such rough terrain.

On arriving at an outcrop, the geologist walks all over it and makes notes on the characteristics of the rocks and on any relationships visible, and samples unusual rocks for laboratory examination. The passenger's side of our vehicles had a piece of plywood mounted on the dashboard as a small desk, and when we had finished looking at an outcrop, we used this desk to draw geological boundaries and faults between different rock units directly on to the air photos with red grease pencil, adding symbols to denote the orientation of bedding, foliation, layering, etc.

Geologists look for several things as they walk around. The simplest is an area of apparently uniform rock type: here we look at and describe the various characteristics of the rock, such as its grain size, texture (the shapes of the grains and how they are attached to each other), and structures such as cross-bedding, which can tell us if a sedimentary rock is right way up or upside down. We also look at layering or foliation[3] if it's an igneous or metamorphic rock, and identify the rock's mineral composition. We measure the orientation—angle of dip below the horizontal, and bearing of the strike (or trend) with respect to due north—of sedimentary bedding, or of igneous or metamorphic layering or foliation. We take a good look around an exposure to make sure that what appeared uniform on the air photo really is uniform when looked at close up, and to be sure that we have seen all there is to see. Two of the best bits of advice my first party leader, David Forman, gave me were, 'Always look on the other side of the hill' and 'In field notes, always write down the evidence for what you conclude.'

Complex outcrops usually show some discontinuity or change on the air photo. A striped appearance on the air photo usually indicates interlayering of several different rock types, and each must be described as outlined above. A sharp discordance on the air photo can be caused by a fault—an extensive surface kilometres in extent where large rock masses have slid past each other—or unconformity, where older rocks have been tilted, eroded, and later covered by younger sedimentary rocks. A finely linear photo pattern in an otherwise uniform rock can be caused by the parallel arrangement of flaky or spindly grains that make up the rock. So we visit these exposures to find out the cause of the complexity.

Some areas are difficult to map even when we are standing right on top of them. The most common difficulty arises where the rocks are strongly weathered. This is because Australia almost completely escaped the Earth's most recent ice age, which ended about 11,000 years ago. Ice covered large areas of the northern hemisphere, and as it advanced it scraped away weathered rock off the underlying bedrock, which later appeared as extensive areas of clean fresh rock when the ice retreated. We can see these young, fresh exposures in places like Scotland, Canada, and the Alps. In contrast, much of Australia is covered in a mantle of weathered rock, where almost all the minerals that make up fresh hard bedrock, such as feldspar, mica, etc., have been altered to clay. In these areas, it is difficult to recognise the hard rock that existed before weathering, so one cannot decipher, from ground observations alone, the overall history of the region. Here, we must rely on other information, such as drilling results or geophysical data, if they exist, to find out the nature of the bedrock. Thick soil cover and vegetation can also conceal bedrock and cause similar difficulties for hard rock mapping.

3 Foliated rocks contain flaky or needle-shaped mineral grains arranged parallel to each other, making the rock easy to split.

Aerial photograph of part of the Davenport Range, showing large-scale folds and faults in hard sandstone (ridges) and softer sandstone and shale (valleys) on northeastern flank of Skinner Pound. The prominent fracture extending northwest from near the centre is one such fault. Field of view 5.6 km by 3.7 km. North is at the top. The aerial photograph is Crown Copyright, and is reproduced by permission of the Chief Executive Officer, Geoscience Australia.

A different sort of difficulty can arise in the geologist's own mind. Geologists bear in mind several possible explanations for the relationships and origins of rock that they see. Hence, many fields geologists look for the critical pieces of evidence that fit one (or a few) of these explanations. This is the method of 'multiple working hypotheses'. This method does not suit some geologists (such as myself), for it seems to me that it can lead to prejudging the issue—we approach the rocks with pre-conceived ideas in mind, and naturally concentrate on looking for the evidence that bears on those ideas. We may blind ourselves to other evidence that bears on other ideas.

An illustration of this arose in the Mount Isa work. In 1987, Monash University students and staff members worked out of our base camp at Lake Julius, and when we returned to camp from our weekly traverse, we used to listen to their discussion on the results of their week's work. Every week, the interpretation of their mapping changed—the previous week's story, so sure and positive then, was discarded, and a new and different interpretation adoped. In the end they found the answer that fitted all the facts, but they spent a lot of time going down blind alleys. I preferred to note all the evidence in a wholly unprejudiced way, and then see which working hypothesis fitted the data when synthesising the area's geological evolution.

Yet another difficulty can be the remoteness of an area. The time and expense necessary to visit exposures may be judged too much for the potential result. One example was the Petermann Ranges 1:250 000 map sheet, in the southwest corner of the Northern Territory. This sheet was mapped by BMR in 1963 as part of the Australian Government's Accelerated Oil Search program, but previous reconnaissance suggested it was made up of rocks without petroleum potential. It had however, following mapping of the Bloods Range sheet in 1962 (chapter 2) become critical to a radical and complex interpretation by David Forman of the structure and evolution of the southern margin of the whole Amadeus Basin. So its northern and southern parts were mapped at the normal 1:250 000 scale, as they were within reach of roads and tracks. The central part of the Sheet area is, however, a 'sea' several thousand square kilometres in extent of sand dunes that surround small hills and ridges of bedrock. These were not visited for another 30 years, when new roads had been constructed across the area, and the Northern Territory Geological Survey produced a second edition of the map showing the new results in 1999.

On returning to base camp from traverse, the geologist spent much of the next day delineating boundaries between different rock types, fault lines, and symbols showing the inclination of layers or foliation, on the air photos from the week's traverse just finished, before giving the completed photos to the field draftsman—which most parties had—and then preparing the next week's work. The draftsman traced the line work on to photo-scale compilation sheets. These were large sheets of translucent film that displayed topographic detail such as water courses, crests of sand dunes, tracks, or in featureless country, such as the sand plain in the Tanami Desert, photo control points, all in their correct positions, with the radial distortion in the air photos removed. By the end of the field season or a few weeks later, the draftsman had a set of completed compilation sheets covering a map area. These were photographically reduced to 1:250 000 scale, taped together, and the whole map traced by hand on to a single transparent sheet of film. After the addition of a map legend, plus the names of the authors, date of publication, and map title, about 50 uncoloured copies of each map were printed and put in a pocket at the back of a roneoed report.[4] The legend included a list of rock units and their description in order from the oldest at the bottom of the list to the youngest at the top, as well as structural and topographic symbols and their meanings, and sundry other items. The report also included photos of significant outcrops in the area described, and ammonia dyeline prints of line diagrams such as a locality map and any diagrams necessary to explain geological ideas. These *Records* were a rapid way of making new information available as quickly as possible, and BMR produced many hundreds of them in the 1960s and 1970s.

Throughout the 1:250 000 mapping program, it was clear to geologists primarily interested in bedrock or 'hard' rocks (i.e., rocks below superficial cover) that the black-and-white air photos were of insufficient clarity and scale to show the details of complex areas. By 1967 BMR became convinced of this and flew the first experimental colour air photos in Australia over the area around Napperby homestead, northwest of Alice Springs, where Allan Wells had a field mapping party.

4 To produce a roneoed report, text was typed on to a foolscap sheet – about 13 inches by 8 – of opaque plastic film, so that each letter cut partly through the film. Each completed sheet of film was wound around a drum in a duplicating machine, and then a single revolution of the drum against a page of blank paper accompanied by ink passed through the film, printed the text on to the paper. Fifty rotations produced fifty pages. Fixing any typing errors on the film before duplicating was even more cumbersome. The typist painted a red solution of some sort of plastic dissolved in an organic solvent over the error, and when it dried, a new patch of (red) film was supposedly in place, ready to be typed over with the correct word. Messy smudges sometimes resulted. Photocopy machines fortunately made duplicating machines obsolete.

The first prints showed low contrast, pale, pinkish orange circular areas, each one inside a black square. To be candid, they looked very funny to old hand geologists, and there was some laughter at the time. But the technique improved rapidly, and in 1969 Geoff Derrick's party mapped the entire Cloncurry 1:100 000 sheet using colour air photos at 1:25 000 scale.

The new-generation colour air photos revealed so much complexity in the 'hard' rocks that the speed of 1:100 000 mapping slowed considerably compared to 1:250 000 scale mapping. As a recompense, the accompanying reports became much larger—some were several hundred pages long—and took longer to produce, sometimes appearing several years after mapping of an area had begun.

In the 1970s, the first images taken by Earth-orbiting satellites began to appear. These were of very large areas, up to several hundred square kilometres in each frame—too poor in detail for geological mapping. As contrast and sharpness improved, the images became useful for displaying large areas in their entirety, but it was not until the 1990s, when calibration and interpretation of the images had advanced to the stage of identifying concentrations of specific minerals, that satellite images could be used for detailed studies and mapping. Areas of hard rock are commonly altered to specific minerals in a halo up to several kilometres in extent around ore deposits, and so these minerals can be used to locate areas worth prospecting where the actual orebody is not obvious, or concealed.

APPENDIX II

AREAS MAPPED

Joined BMR 2 Apr 62		
May 62–Oct 62	Amadeus Basin	Fieldwork in Bloods Range and parts of Lake Amadeus, Scott, and Petermann Ranges
Oct 62–Apr 63	Amadeus Basin	Producing Bloods Range map and report
May 63–Oct 63	Amadeus Basin	Fieldwork in Kulgera, Finke, and part of Ayers Rock area
Oct 63–Apr 64	Amadeus Basin	Producing maps and report
May 64–Oct 64	Amadeus Basin	Fieldwork in Hale River, McDills, and parts of Alice Springs, Illogwa Creek and Finke
Oct 64–Aug 65	Amadeus Basin	Producing maps and reports
Resigned from BMR		
Sep 65		Began post-graduate study in USA
Jun 66–Aug 66	Western Idaho	Fieldwork for thesis
Jun 67–Aug 67	Arltunga area, NT	Fieldwork for thesis
Jun 68–Jul 68	Ringwood drilling	Logging core and cuttings
Aug 68–Jan 69	Arltunga area, NT	Fieldwork for thesis
Jan 69–Oct 70		Laboratory work and writing PhD thesis
Rejoined BMR Jan 71		
Jan 71–May 71	Arunta Region	Laboratory and office work
May 71–Sep 71	Arunta Region	Fieldwork in Fergusson and Reynolds Ranges
Oct 71–May 72	Arunta Region	Laboratory and office work
May 72–Oct 72	Arunta Region	Fieldwork in Reynolds Range and Mount Theo areas
Nov 72–Jul 74	Arunta Region	Laboratory and office work
Jul 74–Sep 74	Arunta Region	Fieldwork, Denison and Mount Peake
Oct 74–May 75	Arunta Region	Laboratory and office work
May 75–Jun 75	Arunta Region	Field inspections in Musgrave, Strangways, and Reynolds Ranges
Jun 75–Jul 76	Arunta Region	Office work

Jul 76–Sep 76	Amadeus Basin, Arunta Region	Fieldwork and International Geological Congress excursions
Sep 76–Jun 77	Arunta Region	Office work
Jun 77–Jul 77	Qld and NT	Field inspections of BMR project areas
Jul 77–Apr 79	Arunta Region	Map production and writing up work
Apr 79–Jun 79	Yilgarn Region, WA	Fieldwork in Youanmi area
Jun 79–Jul 79	Arunta Region	Office work, writing up
Jul 79–Sep 79	Yilgarn Region, WA	Fieldwork in Youanmi area
Sep 79	Yilgarn Region	Office work, Perth, GSWA
Oct 79–Jul 80	Yilgarn and Arunta regions	Office work: producing Youanmi and Arunta Region maps
Jul 80–Feb 81	Yilgarn Region	Writing up Youanmi work
Feb 81–May 81	Yilgarn and Arunta regions	Writing up Youanmi and Arunta region work
May 81–Oct 81	Davenport Region	Fieldwork in Kurundi area
Oct 81–Feb 82	Arunta Region Davenport Region	Writing up office work
Feb 82	Saudi Arabia	Exchange visit to Saudi field parties
Mar 82–Jun 82	Davenport Region	Laboratory and office work and writing up
Jun 82–Jul 82	Davenport Region	Fieldwork in Kurundi area
Aug 82	NT	Exchange visit by Saudi Arabian geologists
Aug 82–Sep 82	Switzerland	Conference and field trips
Sep 82–Oct 82	Davenport Region	Fieldwork in Kurundi and Ooradidgee areas
Oct 82–Jun 84	Davenport Region, Arunta Region	Map production and writing up
Jun 84–Jul 84	Mount Isa Region	Fieldwork in Leichhardt River area
Aug 84	Amadeus Basin	Fieldwork in Undoolya and Fergusson Range
Sep 84–Jun 85	Davenport Region Amadeus Basin	Writing up Map production
Jun 85–Aug 85	Mount Isa Region	Fieldwork in Leichhardt River and Lawn Hill
Sep 85–May 86	Mount Isa Region Davenport Region	Office work and writing up
May 86–Jul 86	Mount Isa Region	Fieldwork in Mount Frosty area
Aug 86–May 87	Mount Isa Region Amadeus Basin	Office work and writing up Map production
May 87–Jul 87	Mount Isa Region	Fieldwork in Deighton area
Jul 87–May 88	Mount Isa Region Amadeus Basin	Office work, writing up Map and cross-section production
May 88–Jun 88	Mount Isa Region	Fieldwork in Deighton and Mary Kathleen areas
Jun 88–Jul 88	Mount Isa Region	Office work
Jul 88–Aug 88	Mount Isa Region	Fieldwork in Ballara—Wee MacGregor area
Aug 88–Jun 89	Mount Isa Region Amadeus Basin	Office work Writing up, map production

Jun 89–Aug 89	USA and Europe	International Geological Congress and rafting in Grand Canyon, rail travel in Europe
Sep 89–Jun 90	Mount Isa Region Amadeus Basin	Map production and editing *Bulletin* Map production
Jul 90–Sep 90	Musgrave Region	Fieldwork in Tomkinson Ranges
Sep 90–Jun 91	Mount Isa Region Amadeus Basin Musgrave Region	Map production, writing and editing *Bulletin* Workshop and writing up Laboratory work
Jul 91–Sep 91	Musgrave Region	Fieldwork in Tomkinson Ranges and Bates area
Sep 91–Jul 92	Mount Isa Region	Map production, writing and editing *Bulletin*
Jul 92–Aug 92	Musgrave Region	Fieldwork in Umbeara area
Aug 92–Oct 93	Mount Isa Region Musgrave Region	Finalised *Bulletin* Laboratory and office work, map production, writing up
Oct 93–May 94	Eastern Goldfields Musgrave Region	Administration Laboratory and office work, map production, writing up
May 94–Aug 94	Eastern Goldfields	Fieldwork in Duketon and Sir Samuel areas
Aug 94–May 95	Eastern Goldfields Musgrave Region	Laboratory and office work, map production Map production, writing up
Jun 95–Sep 95	Eastern Goldfields	Fieldwork in Lake Violet and Yeelirrie areas
Oct 95– Apr 96	Eastern Goldfields Musgrave Region	Laboratory and office work, map production Map production, writing up
Apr 96–Jun 96	Eastern Goldfields	Fieldwork in Lake Violet area
Jun 96–Sep 97	Eastern Goldfields Musgrave Region	Office work, map production, writing up Writing up
Oct 97	Eastern Goldfields	Kal '97 Conference and field checks
Oct 97–Sep 98	Eastern Goldfields Musgrave Region	Office work, map production, writing up Writing up
Sep 98–Nov 98	Eastern Goldfields	Fieldwork in Leonora area
Nov 98–Mar 99	Eastern Goldfields	Map production, writing up
Mar 99	Eastern Goldfields	Fieldwork in Leonora area (aborted by Cyclone Vance)
Mar 99–May 99	Eastern Goldfields	Laboratory work, map production, writing up
Jun 99–Jul 99	USA	Travel
Aug 99	Eastern Goldfields	Map production, writing up
Sep 99–Oct 99	Eastern Goldfields	Fieldwork in Leonora area
Nov 99	Eastern Goldfields	Office work, map production, writing up
Retired from Geoscience Australia 30 Nov 99		
Dec 99–Jul 2000	Eastern Goldfields	Writing up, map production, editing
Aug 00–Oct 00	Arunta Region	Fieldwork (as field hand)
Oct 00–Dec 00	Eastern Goldfields	Map production

Area mapped: About 83,000 square kilometres, or approximately 1.1 per cent of Australia.

GLOSSARY

Alluvium: sand, gravel, and mud deposited in a stream channel or floodplain.

Amphibolite: tough dark metamorphic rock formed by deep-seated heating and compression of a parent rock containing substantial amounts of iron and magnesium.

Anhydrite: calcium sulphate formed by heating and dehydration of gypsum during burial.

Bedrock: good hard unweathered (or fresh) rock below weathered rock and its redistributed detritus.

Benchmark: fixed mark of known height set up by a surveyor.

Deadman: large object, usually a log, buried in the ground to serve as a winch cable attachment point when a vehicle needs to winch itself out of trouble.

Bulldust: very fine dust of any composition that moves like a liquid when disturbed. Cattle enjoy kicking it up.

Dolomite: calcium and magnesium carbonate $CaMg(CO_3)_2$ that forms thick sedimentary rock layers at the bottom of shallow seas. The crushed rock is often added to garden soil to reduce soil acidity.

Evaporites: sedimentary rocks formed by crystallisation of various minerals, e.g., salt and gypsum, when sea or lake water dries up.

Fault: a fracture along which rocks have moved.

Fold: a bend in layered or foliated rocks.

Geode: hard ovoid mass of concentric mineral shells, often amethyst, deposited in a large cavity in silica-rich volcanic rocks. Geodes commonly lie around on the ground after weathering and erosion have removed the softer volcanic host rock.

Grain: a mineral or rock particle with diameter less than a few millimetres, generally lacking well formed crystal faces.

Gypsum: calcium sulphate, $CaSO_4.2H_2O$, used for making plasterboard or gyprock.

Igneous rocks: rocks formed by cooling and solidifying of molten rock, or magma.

Limestone: sedimentary rock formed by deposition of particles of calcium carbonate, $CaCO_3$.

Mafic rocks: rocks composed of roughly equal amounts of magnesium-iron silicates and sodium-calcium-aluminium silicates.

Metamorphic rocks: rocks formed by growth of new minerals from old minerals without melting; the new minerals commonly show a preferred orientation.

Sedimentary rocks: rocks formed by the deposition of mineral grains in air or water, followed by cementation of the grains together.

Silica: the oxide of silicon, SiO_2, commonly found as quartz grains in beach and dune sand, as agate and opal, and in sandstone.

Siltstone: a sedimentary rock made of particles intermediate in size between sand and clay.

Tectonic province: a region that has similar ages of rock formation and similar deformational structures, such as folds, faults, and foliation, that differ from those of adjacent regions.

Ultramafic rocks: rocks composed almost wholly of magnesium-iron silicates.

Volcanic rocks: rocks solidified from lava poured or blasted out by a volcano.